ENVIRONMENTAL REMEDIATION TECHNOLOGIES, REGULATIONS AND SAFETY

ECOLOGICAL AND BIOGEOCHEMICAL CYCLING IN IMPACTED POLAR ECOSYSTEMS

Environmental Remediation Technologies, Regulations and Safety

Additional books in this series can be found on Nova's website under the Series tab.

Additional e-books in this series can be found on Nova's website under the eBooks tab.

ENVIRONMENTAL REMEDIATION TECHNOLOGIES, REGULATIONS AND SAFETY

ECOLOGICAL AND BIOGEOCHEMICAL CYCLING IN IMPACTED POLAR ECOSYSTEMS

VLADIMIR N. BASHKIN

EDITOR

nova science publishers
New York

Library of Congress Cataloging-in-Publication Data

ISBN: 978-1-53612-081-3

Published by Nova Science Publishers, Inc. † New York

CONTENTS

INTRODUCTION

ECOLOGICAL AND BIOGEOCHEMICAL CYCLING IN IMPACTED POLAR ECOSYSTEMS

Prof., Dr. Vladimir N. Bashkin[*,1,2]

[1]Institute of Physicochemical and Biological Problems of Soil Sciences of Russian Academy of Sciences, Pushchino, Russian Federation

[2]Institute of Natural Gases and Gas Technologies, Gazprom VNIIGAZ LLC, Razvilka, Russian Federation

Keywords: ecological and biogeochemical cycling, polar ecosystems, gas industry impact, biogeochemical technologies, risk management

ABSTRACT

This paper presents the editor's introductory words for the set of articles devoted to ecological and biogeochemical cycling in impacted polar ecosystems. Special attention is provided to developing

* Address correspondence to this author at Institute of physicochemical and biological problems of soil science RAS: Prof., Dr (biol) Vladimir Bashkin, Principal researcher, Pushchino, Moscow region, 142292 Russia. Tel: +7 916 860-2038, Fax: +7 4967 33-0595, e-mail – Bashkin@issp.serpukhov.su

biogeochemical technologies for managing pollution in these ecosystems as well as the relevant risk management.

The production of hydrocarbons in the Far North, as with other types of economic activity, is accompanied by changes in the natural environment. The scope and nature of these changes depend on the intensity and specifics of ongoing impacts, determining the local and regional pollution levels, this or that industrial (or secondary) geo-ecological situation and the relevant geo-ecological conditions in areas of development of deposits. For example, from the resilience of soils to thaw, as defined by species of vegetation, and in turn, by emissions of nitrogen species, which are both very important nutrients and pollutants, depends on the successful exploitation of technological equipment and infrastructure of gas and oil industry.

These criteria are largely determined by the stability of the biogeochemical structure of the Arctic ecosystems in the impacted zones of the enterprises of the gas and oil industry. Specific emissions of different gaseous compounds during extraction, preparation, transportation and processing of natural gas and associated components are the main reason of changes, which occur in biogeochemical cycles of carbon and nitrogen. This means, accordingly, that the anthropogenic load of these elements should be limited in the ranges of natural variations of different links of biogeochemical trophic chains of these cycles without irreversible changes.

Thus, the study of ecological and biogeochemical indicators in gas industry impacted polar ecosystems in the Northern regions, including footprints of climate changes, the simulation of geoecological and geopolitical risk processes and the relevant challenges allows us to simulate the behavior of key elements (nitrogen, carbon) in biogeochemical cycles. This, in turn, can be used in the development and application of biogeochemical technologies, for example, for recultivation of disturbed and polluted areas, calculation of geo-ecological risk indicators and for developing techniques to manage this risk in different geo-ecological situations owing to environmental pollution.

This edited volume is accordingly devoted to one of the hottest topics of modern ecological and environmental sciences, i.e., to the alterations in ecological and biogeochemical cycling in impacted polar land and water ecosystems where at present the significant part of natural gas and oil production is developing both in Russia and in many other countries and in the nearest future these will be in the great progress. Being founded on the field monitoring of various parameters of biogeochemical cycles of polar ecosystems and relevant changes in ecological indexes and indicators, the authors suggest new biogeochemical technologies for assessing and managing environmental and geoenvironmental risks in severe climate conditions.

Accordingly, I would like to emphasize my main idea of writing the given volume - this is to share our experience in estimating gas industry impacts on ecological and biogeochemical cycling in tundra ecosystems and it seems to me that the Russian experience is quite unique due to the gas production development in this area for a long time. Of course, the authors of this volume, both experienced and young researchers, are from different companies and institutions of Russia, such as Gazprom R&D Institute of Natural Gases and Gas technologies, Institute of Physicochemical and Biological Problems of Soil Sciences RAS, Institute of Basic Biological Problems RAS, Northern (Arctic) Federal University named after M.V. Lomonosov, Institute of Soil Science and Agrochemistry of RAS Siberian branch, and I appreciate greatly their contributions to the given volume. Furthermore, I am very thankful to my international colleagues from Sweden for presenting new ideas as how to estimate the north ecosystem critical loads accounting for the vegetation indexes.

Finally, I think we are in position to present the comprehensive picture of suggested topic.

Hopefully, this issue is meant for a wide range of scientists, specialists and practitioners in ecological science, biogeochemistry and energy sector and may be quite useful to undergraduate and postgraduate students and masters, who study the fundamental principles of ecosystem sustainability at the anthropogenic impact.

PART I. ECOLOGICAL AND BIOGEOCHEMICAL CYCLING

In: Ecological and Biogeochemical Cycling … ISBN: 978-1-53612-081-3
Editor: Vladimir N. Bashkin

Chapter 1

ECOLOGICAL AND BIOGEOCHEMICAL FOOTPRINTS OF POSSIBLE CLIMATE ALTERATIONS IN POLAR AREAS

***Vladimir N. Bashkin*[1,2] *and Rauf V. Galiulin*[3]**
[1]Institute of Physicochemical and Biological Problems of Soil Science of Russian Academy of Scences, Pushchino, Russian Federation
[2]Gazprom VNIIGAZ LLC, Razvilka, Russian Federation
[3]Institute of Basic Biological Problems of Russian Academy of Sciences, Pushchino, Russian Federation

ABSTRACT

In this paper ecological and biogeochemical footprints are shown for possible climate alterations in polar areas of the Yamal Peninsula, the Russian Federation. The relative content of soil nitrogen and carbon fractions of soil organic matter can be served as ecological footprints of regional climate changes. These soil organic matter values may grow during the cyclic periods of regional temperature increase and may decline in the relevant cold periods. The temperature ranking may serve as the biogeochemical footprints of regional climate changes as well. The simulating climate changes in the Yamal Peninsula indicated the modern predominant trends to relevant cooling periods.

Keywords: gas industry impacted ecosystems, polar region, climate deviation modeling, ecological footprints, biogeochemical footprints

INTRODUCTION

Depending on climate, geological bedrock, soil and vegetation cover, hydrology, and topography, the global ecosystem is divided into various ecoregions and ecozones (see for example, Bailey 1998). In all these structural units of the biosphere, the spatial organization of the global ecosystem is manifested in the form of characteristic features of the biogeochemical cycles of different elements. This applies equally to the polar ecosystems.

Biogeochemical cycles in different ecosystems are largely determined by the biota, especially the processes of primary productivity and decomposition of dying organic matter. Biogeochemical cycles represent the cumulative representation of the circulation of substances between the various components of the biosphere – soils, surface and groundwater, bottom sediments, biota and atmosphere.

Soil, geobotanical, geological, and overall ecosystem regionalization is the basis of biogeochemical zoning (Glazovskaya 1997; 2002; Fortescue 1980; Glazovskaya 1984; 1997; 2002; Bazilevich 1993; Ermakov 1993; Bashkin & Bailey 1995; Kasimov 1995; Perelman & Kasimov 1999; Kasimov & Clige 2006; Bashkin 2002; 2004; 2006; 2009; 2014; 2016; Bashkin & Kasimov 2004). The combination of these types of regionalization with the quantitative indicators of biological, geochemical and hydrochemical cycle gives the possibility to calculate the speed of the biogeochemical cycle.

Tundra ecosystems (as indeed all of the high-latitude and high-altitude ecosystems in general) are the most vulnerable to all sorts of both anthropogenic influences and conditionally natural climate change.

However, the interaction of anthropogenic effects (e.g., atmospheric contamination by pollutants) with a global change may cause different consequences, influencing all components of these ecosystems, their functioning in general and to change their biospheric functions.

Ecological Footprints of Climate Changes in Polar Ecosystems

It is established that the increase in soil temperature leads to increased microbial immobilization of nitrogen and increased heterotrophic respiration, thus reducing net primary productivity of ecosystems (Chapin et al. 1995a; Vourlitis & Oechel 1999; Bashkin 2008). Nevertheless, after 3-10 years of warming increased rate of digestion results in an increase in the rate of absorption of nitrogen by plants and increased net primary productivity (Shaver et al. 2000). However, it was made few attempts to study the binding mechanisms for the observed adaptations of the processes of transformation of nitrogen.

At the same time the beginning of a change in trend from warming to cooling since the early 2000s (Bashkin 2016) shows a number of experimental data, in particular, the increase in the permafrost zone in the area of gas condensate fields where the relevant monitoring is conducting. This is accompanied by preservation of the existing natural environment, including biogeochemical cycles of elements.

In the result of permafrost and the active cryogenic processes the Arctic ecosystems represent a wide range of surface types: circles, polygons, nets, steps, stripes (Washburn 1956; Vasilevskaya 1996). Because the soil associated with these microsections often differ in abiotic and biotic factors, soil properties and vegetation cover can change dramatically over very short distances. Among the most common surface structures in permafrost areas are earthen mounds (bumps), which represent the "primary network" in accordance with the classification of patterned structures of the Earth's surface by Washburn (1956). Bumpy tundra is characterized by small earthen bumps and intermediate depressions and organized with marked regularity. However, the size of the bumps can vary widely. Earthen bumps should be distinguished from peat ones formed in bogs consisting of peat and mosses, and the bogs formed mostly by cottongrass, sedge and some shrubs.

The given bumps are typical type of surface the polar region. Arctic ecosystems are generally characterized by low available nutrients due to the cold and wet conditions and, thus, the low net primary production (Manakov 1972; McKane et al. 1997; Wookey 2002; Dobrovolsky 1994; 2003; Bashkin et al. 2012). However, spatial differences in microtopography and associated differences in physicochemical soil properties, together with a variety of functional characteristics of plant species, dominant on the hills and in depression areas probably also are associated with a remarked mosaic of cycling and availability of nutrients.

Considering global climate change and its impact on Arctic ecosystems, it has been hypothesized that warming will lead to significant vegetation change: to reduce the biomass of mosses (Chapin et al. 1995b) and a significant increase in the proliferation of shrubs (Sturm et al. 2001). In addition, warming can cause changes in the intensity of cryoturbation processes, giving, thus, the change in the result of transformation processes in Arctic soils (Goryachkin et al. 1999). Taken together, these changes in the environment can significantly affect the structure and function of bumpy tundra, but the magnitude and sequence of such changes is still unclear, since our knowledge of the patchiness of this complex system is very limited. Because Arctic ecosystems store large stocks of soil carbon (Vasilevskaya 1998, Bashkin 2014) changes in biogeochemical cycles can also be very important when considering a reverse correlation with climate warming. Detailed studies of the nitrogen pool in soils of tundra ecosystems and its heterogeneity in space are the necessary basis to study the effects of elevated temperatures on biogeochemical cycles of nitrogen and carbon.

Nitrogen is the most deficient element in mineral nutrition of plants and soil organisms in tundra ecosystems and, thus, is very important for the formation of net production in Arctic ecosystems (Oechel et al. 1993;

2000; Dobrovolsky 2003). Moreover, Le Dize et al. (2003) demonstrated the role of nitrogen as an important limiting factor controlling the flows of carbon in tundra ecosystems. Therefore, the content of available forms of nitrogen largely determines the functioning of tundra ecosystems as a whole. In turn, the pool size of available nitrogen in the soil (mainly in the form of nitrates, water-soluble and exchangeable ammonium) is determined by the speed of the biogeochemical processes through which these forms of nitrogen are fed in the system soil – plants (mineralization, nitrification, release of fixed ammonium) and spent it (immobilization heterotrophic soil microbiota, assimilation by plants, denitrification, nitrification, non-exchange fixation of ammonium by clay minerals, leaching of nitrate from the soil profile) (Bashkin 2002; 2016; Bashkin et al. 2012).

In gas industry impacted polar ecosystems nitrogen anthropogenic emission in the form of NO_x is the main source of pollutant input. Since the biogeochemical cycle of nitrogen and carbon are closely interrelated, the input of NO_x is changing both nitrogen and carbon turnover. Accordingly, the study of nitrogen and carbon transformations and pools of tundra ecosystems is needed in the aspect of anthropogenic pollution and global climate changes. The ecological footprints of the given changes can be monitored as the alterations of carbon and nitrogen turnover.

Biogeochemical Footprints of Climate Changes

Temperature Rating of Biogeochemical Cycling

In most parts of the Earth's surface the actual length of the various processes that determine the cycling of chemical elements (chemical, biological, microbiological, geochemical, and biogeochemical processes) depends on the duration of the winter period, when the intensity of these processes is significantly reduced or completely suspended for a period of 6-10 months. It is also known that the impact of any pollutants on these processes occurs only during the spring and summer time. The duration of this impact can be characterized by the coefficient of active temperatures, C_t, representing the ratio of the sum of active temperatures > 5°C to the total amount of annual temperatures.

To account for the temperature influence on the process of biological absorption of elements is proposed by using the coefficient of biogeochemical cycles, C_x, which is calculated as the product of the coefficient of inner-ecosystem biological cycle, C_b equal to the ratio of the mass of litter to the mass of the annual litter, and coefficient of active temperatures:

$$C_x = C_b \times C_t.$$

Its use allows taking into account the impact of the period of active temperatures on the rate of biogeochemical cycles in different ecosystems. The coefficient of biogeochemical cycle can be used for correction of the values of inner-ecosystem biological cycle, C_b.

Table 1 shows the combination of soil temperature and biogeochemical conditions in various polar and tundra geographic regions of the Earth.

Table 1. The coefficients inner-ecosystem biological cycle (C_b), active temperatures (C_t) and biogeochemical cycles (C_x) in different soil ecosystem and geographical regions of the Earth

Ecosystems	Main soil FAO types	Geographical regions (Glazovskaya 1978; Bailey, 1998; Bashkin & Kozlov 1999)	C_b	C_t	C_x
Arctic deserts and primitive tundra	Litosols, Regosols	North-American	10.0	0.07	0.7
		Eurasian	10.0	0.06	0.6
Tundra	Cryic Gleysols, Histosols, Humic Podzols	North-American	18.0	0.16	2.8
		Eurasian	18.0	0.15	2.7

There are represented ecosystem types, soil types according to FAO classification, the coefficients of inner-ecosystem biological cycle, C_b, active temperature, C_t, and biogeochemical cycles, C_x.

Table 2. Ranking of temperature mode parameters for assessing the period of active biogeochemical reactions

Rank	Temperature regime	Active temperature coefficient, C_t
1	Arctic	<0.25
2	Boreal	0.26–0.50
3	Sub-Boreal	0.51–0.80
4	Mediterranean	0.81–0.99
5	Sub-Tropical & Tropical	1.00

Table 3. Ranking of indicators of biogeochemical cycle to assess the rates of migration in different ecosystems

Rank	Biogeochemical cycling rank	Coefficient of biogeochemical cycling, C_x
1	Very intensive	<0.4
2	Intensive	0.5–1.4
3	Moderate	1.5–2.5
4	Depressive	2.6–4.9
5	Very depressive	>5.0

To understand these approaches it is necessary to pay attention to the principles of the global regionalization of the applied parameters. The coefficients of active temperatures, ranked in accordance with the main climatic zones are shown in Table 2.

The coefficients of biogeochemical cycles, C_x, for all geographic regions were ranked to determine the type of biogeochemical cycle and these grades are shown in Table 3.

Selected 5 types of biogeochemical cycles: very intense, intense, moderate, depressive and semi-depressive. Intensive and very intensive type of biogeochemical cycle occurs in tropical and desert ecosystems, which are experiencing rapid biological turnover under conditions of high temperature of the growing season. A moderate type of circulation is typical for the steppe, forest-steppe and sub-boreal forest ecosystems, whereas the boreal taiga forests are dominated by depressive type of biogeochemical cycle. Very depressive cycle of the cycle has been in the tundra and Arctic ecosystems.

Thus, the gas industry impacted polar and tundra ecosystems have very depressive type of biogeochemical cycling and any climate changes determine the altering of biogeochemical parameters.

Forecast of Climate Change on the Yamal Peninsula

In the long term, a particular problem can cause the transformation of the frozen strata of the Yamal Peninsula under the influence of global climate change.

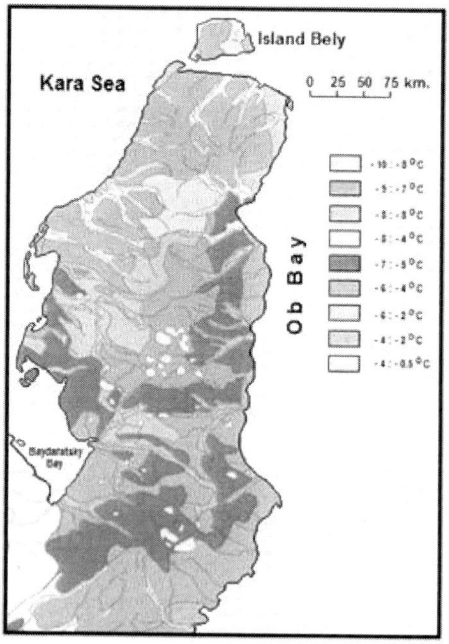

Figure 1. Modern map of average annual values of soil temperature on the Yamal Peninsula.

Figure 1 presents modern map of average annual values of soil temperatures on the Yamal Peninsula, while Table 4 provides the predictive characteristics of the upper horizon of permafrost on this Peninsula at the different trends of changing air temperature.

Table 4. Predicted characteristics of the upper horizon of permafrost on the Yamal Peninsula at the different trends of changing air temperature

<table>
<tr><td rowspan="2">Permafrost composition</td><td rowspan="2">Mean annual ground temperature, °C</td><td colspan="2">Prognoses ground temperature, °C at trends*</td><td colspan="2">Time of ground throwing starting at trends</td><td colspan="2">Throwing depth (m) at trends</td></tr>
<tr><td>0.06°C in year</td><td>-0.03°C in year</td><td>0.06°C in year</td><td>-0.03°C in year</td><td>0.06°C per year</td><td>-0.03°C per year</td></tr>
<tr><td>Sands</td><td rowspan="3">-7.0...-9.0 (modern time)</td><td rowspan="3">-2.0...-4.0</td><td rowspan="3">-5.0...-7.0</td><td rowspan="3" colspan="2">No throwing</td><td rowspan="3">Seasonal, less than 1.5</td><td rowspan="3">Seasonal, less than 1.0</td></tr>
<tr><td>sandy loam, loam</td></tr>
<tr><td>Peats</td></tr>
<tr><td>Sands</td><td rowspan="3">-9.0...-11.0</td><td rowspan="3">-4.0...-2.0</td><td rowspan="3">-6.0...-8.0</td><td rowspan="3">2080-2090</td><td rowspan="3">No throwing</td><td>up to 2.5</td><td rowspan="3">Seasonal, less than 0.8</td></tr>
<tr><td>sandy loam, loam</td><td>up to 2.0</td></tr>
<tr><td>Peats</td><td>up to 1.5</td></tr>
</table>

* 0.06°C in year – before 2000; -0.03°C in year – after 2014

The most significant changes of the depths of seasonal thawing of the soil, and consequently heat-induced throwing could occur in areas with low vegetation ground cover and in its absence (the exposed surface) under a positive temperature trend. Although this scenario does not come true, even if it could not change the conditions of existence of the permafrost zone. In the conditions of negative temperature we can see the trend of permafrost zone conservation, and even its grown (Bashkin 2016).

However, the development of industrial infrastructure without appropriate measures of geo-environmental protection in industrial zones can cause the degradation of the permafrost zone, resulting in the degradation and transformation of permafrost, the formation of thermokarst, heaving and cracking of soils.

The risks and scale of cryogenic processes associated with possible degradation of massive ice, widespread in the Central Yamal Peninsula, as a result of anthropogenic pressures will increase significantly (Fig. 2). Subsidence of the Earth's surface as a result of the gas production on the Yamal fields can cause substantial changes in the hydrography within the territory of the deposits.

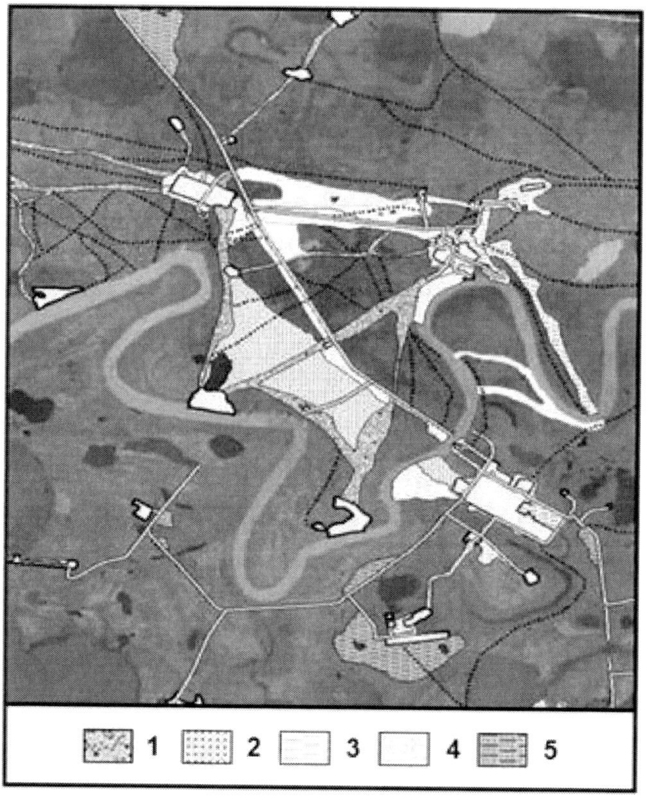

Figure 2. Geodymanic process at gas field area on the Yamal Peninsula: 1 – thermocarst, 2 – thermoerosion, 3 – solifluction, 4 – deflation, 5 – waterlogging.

However, engineering solutions, including those conducted pursuant to long-term monitoring data, are able to provide localization of the negative environmental consequences of this phenomenon.

Planned activities to reduce negative impacts on the natural environment and the protection of the territories and facilities from dangerous cryogenic processes include:

- Development of technological and special events that reduce the negative impact on near-ground atmospheric layer;
- Use of closed water systems, ensuring the prevention of surface water pollution;
- Use of specific technologies, which reduce thermal and mechanical impact on permafrost soils;
- Development of a special delicate area processing technologies;
- Protection of the sites of traditional economic management, archaeological and cultural monuments of the indigenous population.

Conclusion

Thus, ecological and biogeochemical footprints are shown for possible climate alterations in polar areas of the Yamal peninsula, the Russian Federation. The relative content of soil nitrogen and carbon fractions of soil organic matter can be served as ecological footprints of regional climate changes. The microbial consumption of resistant carbon compounds can be explained by changes in the structure of microbial communities dominated by microorganisms with a high temperature optimum, which is capable of utilizing a more sustainable carbon substrate. These soil organic matter values may grow during the cyclic periods of regional temperature increase and may decline in the relevant cold periods. The temperature ranking may serve as the biogeochemical footprints of regional climate changes as well. The simulating climate changes in the Yamal peninsula indicated the modern predominant trends to relevant cooling periods.

Conflict of Interest

The authors confirm that this article content has no conflicts of interest.

Acknowledgments

The authors express their deep appreciation and gratitude to Dr. Pavel A. Barsukov (Institute of Soil Science and Agrochemistry of Siberian Branch of Russian Academy of Sciences) for the support during research and development.

References

Bailey, RG (1998). Ecoregions: Ecosystem Geography of Oceans and Continents. Springer Publishers, New York.

Bashkin, VN (2002). Modern Biogeochemistry. Kluwer Academic Publishers, Dordrech-London-Boston.

Bashkin, VN (2004). Biogeochemistry. Scientific World Publishing House, Moscow.

Bashkin, VN (2006). Modern Biogeochemistry: Environmental Risk Assessment, 2nd edition. Springer Publishers, The Netherlands.

Bashkin, VN (2008). Biogeochemistry. Higher School Publishing House, Moscow.

Bashkin, VN (2009). Modern Biogeochemistry: Environmental Risk Assessment, 2nd edition. CIP Publishers, China – Chinese translation.

Bashkin, VN (2014). Biogeochemistry of Polar Ecosystems in Gas Industry Impacted Zones. Gazprom VNIIGAZ Publishing House, Moscow.

Bashkin, VN (2016). Biogeochemical Technologies for Managing Pollution in Polar Eecosystems. Springer Publishers, Switzerland.

Bashkin, VN, & Bailey, RG (1995). Global map of ecoregions: biogeochemical and soil approaches. *Eurasian Soil Science*, 3, 365-374.

Bashkin, VN, & Kozlov, MY (1999). Biogeochemical approaches to the assessment of East Asian ecosystem sensitivity to acid deposition. Biogeochemistry, 47, 147–165.

Bashkin, VN, & Kasimov, NS (2004). Biogeochemistry. Scientific World Publishing House, Moscow.

Bashkin, VN, Arno, OB, Arabsky, AK, Barsukov, PA, Priputina IV, & Galiulin, RV (2012). Retrospective and Forecast of Geoecological Situation on the Gas Condensate Fields of the Far North. Gazprom VNIIGAZ Publishing House, Moscow.

Bazilevich, NI (1993). Biogeochemistry of Natural Zones of the USSR. Nauka Publishing House, Moscow.

Chapin, FSI, Hobbie, SE, Bret-Harte, MS, & Bonan, GB (1995a). Causes and consequences of plant functional diversity in arctic ecosystems. In: Chapin, FSI and Körner, C. (eds.) Arctic and Alpine Biodiversity: Ecological Studies, 113, Springer-Verlag Publishers, Berlin-Heidelberg 225-237.

Chapin, FSI, Shaver, GR, Giblin, AE, Nadelhoffer, KJ, & Laundre, JA (1995b). Responses of arctic tundra to experimental and observed changes in climate. Ecology, 76, 694-711.

Dobrovolsky, VV (1994). Biogeochemistry of the World's Land. Mir Publishing House, Moscow /CRC Press Publishers, Boca Raton–Ann Arbor–Tokyo–London.

Dobrovolsky, VV (2003). Fundamentals of Biogeochemistry. Academia Publishing House, Moscow.

Ermakov, VV (1993). Biogeochemical mapping of continents. In: Bashkin, VN, Evstafieva, EV, Snakin, VV (Eds). Biogeochemical Fundamentals of Environmental Regulation. Nauka Publishing House, Moscow, pp. 5–24.

Glazovskaya, MA (1978). Soils of the World, Part 1 and 2. MSU Press Publishing House, Moscow.

Glazovskaya, MA (1984). Soils of the World. American Publishing Co Publishers, New Delhi.

Glazovskaya, MA (1997). Methodological Bases of Estimation of Ecological-Geochemical Soil Sustainability to Anthropogenic Influences. MSU Publishing house, Moscow.

Glazovskaya, MA (2002). Geochemical Fundamentals of Typology and Methodology of Research of Natural Landscapes. Ojkumena Publishing House, Smolensk.

Fortescue, JAC (1980). Environmental Geochemistry. A Holistic Approach. Springer-Verlag Publishers, Berlin.

Gregorich, EG, Liang, BC, Drury, CF, Mackenzie, AF, & McGill, WB (2000). Elucidation of the source and turnover of water soluble and microbial biomass carbon in agricultural soil. *Soil Biology and Biochemistry,* 32, 581-7.

Kasimov, NS (1995). Ecogeochemistry of Urban Landscapes. MSU Press Publishing House, Moscow.

Kasimov, NS, & Klige, RK (2006). Modern Global Changes of Natural Environment, 1, Nauka Publishing House, Moscow.

Le Dize, S, Kwiatkowski, BL, Rastetter, EB, Hope, A, Hobbie, JE, Stow, D, & Daeschner, S (2003). Modelling biogeochemical responses of tundra ecosystems to temporal sand spatial variations in climate in the Kuparuk River Basin (Alaska). *J. Geophys. Res.,* 108(D2), doi:10. 1029/2001JD000960.

Manakov, KN (1972). The Biological Cycle in Arctic and Tundra Ecosystems. Nauka Publishing House, Moscow.

McKane, RB, Rastetter, EB, Shaver, GR, Nadelhoffer, KJ, Giblin, AE, Laundre, JA, & Chapin, FS (1997). Climatic effects on tundra carbon storage inferred from experimental data and a model. *Ecology*, 78(4), 1170-1187.

Oechel, WC, Hastings, SJ, Vourlitis, GL, Jenkins, M, Riechers, G, & Grulke, N (1993). Recent change of arctic tundra ecosystems from a net carbon dioxide sink to a source. *Nature*, 361, 520-3.

Oechel, WC, Vourlitis, GL, Hastings, SJ, Zulueta, RC, Hinzman, L, & Kane, D (2000). Acclimation of ecosystem CO_2 exchange in the Alaskan Arctic in response to decadal climate warming. *Nature,* 406, 978-981.

Shaver, GR, Canadell, J, Chapin, FS, Gurevitch, J, Harte, J, Henry, G, Ineson, P, Jonasson, S, Melillo, J, Pitelka, L, & Rustad, L (2000). Global warming and terrestrial ecosystems: A conceptual framework for analysis. *Bioscience,* 50, 871-882.

Sturm, M, Racine, C, & Tape, K (2001). Climate change - increasing shrub abundance in the Arctic. *Nature*, 411, 546-7.

Vasilevskaya, VD (1996). Assessment of the sustainability of tundra permafrost-affected soils to anthropogenic influences. Bulletin of Moscow University, Ser. 17, *Soil Science*, 1, 27-35.

Vourlitis, GL, & Oechel, WC (1999). Eddy covariance measurements of CO_2 and energy fluxes of an Alaskan tussock tundra ecosystem. *Ecology,* 80, 686-701.

Washburn, AL (1956). Classification of patterned ground and review of suggested origins. *Geological Society of America Bulletin*, 67, 170–188.

Wookey, PA (2002). Tundra. Encyclopaedia of global environmental change. In: Mooney, HA, & Canadell JS (Eds) The Earth System: Biological and Ecological Dimensions of Global Environmental Change, J Chichester, John Wiley & Sons Publishers, Ltd., NY, 2, 593-602.

In: Ecological and Biogeochemical Cycling ... ISBN: 978-1-53612-081-3
Editor: Vladimir N. Bashkin

Chapter 2

COMPARATIVE ASSESSMENT OF CARBON AND NITROGEN POOLS IN TUNDRA ECOSYSTEM SOILS

Vladimir N. Bashkin[1,2], Rauf V. Galiulin[3] and Pavel A. Barsukov[4]

[1]Institute of Physicochemical and Biological Problems of Soil Science of Russian Academy of Sciences, Pushchino, Russian Federation
[2]Gazprom VNIIGAZ LLC, Razvilka, Russian Federation
[3]Institute of Basic Biological Problems of Russian Academy of Sciences, Pushchino, Russian Federation
[4]Institute of Soil Science and Agrochemistry of Siberian Branch of Russian Academy of Sciences, Novosibirsk, Russian Federation

ABSTRACT

The given paper was aimed to establish the spatial regularities of the heterogeneous distribution of reserves of different forms of nitrogen and carbon in soils of tundra ecosystems depending on the meso- and microrelief and depth in the soil profile. It was shown that at small scale, the temperature, of course, controls the dynamics of nitrogen in the Arctic ecosystems, which is proved by the higher mean values of the pools of nitrogen and the speed of the cycle in southern tundra compared to the typical. Soil moisture (or the effect of meso-relief) affects the mode of nitrogen in only the soil between bumps along the gradient from temperate-humid to extra-humid conditions. However, in larger scale, the effect of microrelief and the closely related differentiation of vegetation covers the effects of temperature and soil moisture, and acts as the main controlling factor of the nitrogen fund in tundra soils due to the redistribution of litter and the lateral introduction of dissolved organic and mineral substrates in depression zone. The mosaic distribution of nitrogen of the zonal background soils and its rate of spatial mineralization is enhanced by cryoturbation processes that result in party leaching of organo-accumulative horizon with high nitrogen content into the soil depth in conditions of lower temperatures and, accordingly, reduced activity of the microbiota.

Keywords: carbon and nitrogen pools, nitrogen mineral form, tundra soils, monitoring

INTRODUCTION

The functioning and resilience of natural systems in the invasion of anthropogenic fluxes depend on the degree of safety and activity of organic substances and, therefore, stability of the existing biogeochemical processes (Bashkin 1987; 2004; 2006; Moskovchenko 2010). Tundra ecosystems (as indeed all of the high-latitude and high-altitude ecosystems in general) are the most vulnerable to all sorts of anthropogenic influences and natural climate changes.

However, the interaction of anthropogenic effects (e.g., atmospheric contamination by gas industry pollutants) with the global changes may cause different consequences, influencing all components of these ecosystems, their functioning in general and to change their functions in the biosphere.

Nitrogen is the polyfunctional element being the most effective nutrient and serious pollutant. In various places and in particular in polar areas, N is permanently deficient in mineral nutrition of plants and soil organisms in tundra ecosystems and, thus, is very important for the formation of net production (Oechel et al. 2000; Bashkin et al. 2012). Moreover, in the work of Le Dize's et al. (2003) the role of nitrogen was shown as an important limiting factor controlling the flows of carbon in tundra ecosystems. Therefore, the content of available forms of nitrogen, largely determines the functioning of tundra ecosystems as a whole. In turn, the pool size of available nitrogen in the soil (mainly in the form of nitrates, water-soluble and exchangeable ammonium) is determined by the speed of the processes through which these forms of nitrogen are fed in the system soil - plants (mineralization, nitrification, release of fixed ammonium) and spent it (immobilization heterotrophic soil microbiota, assimilation by plants, denitrification, nitrification, non-exchange fixation of ammonium by clay minerals, leaching of nitrate from the soil profile). Furthermore, additional input of anthropogenic nitrogen as emitted NO_x may change not only nitrogen turnover but in the significant degree the cycle of carbon and other biophilic elements.

For understanding the anthropogenically induced changes in the Artic areas we need the knowledge on the background parameters of nitrogen and carbon cycling in tundra ecosystems and accordingly the comparative assessment of the given element pools is of great importance.

Soil Samples Description

Basic soil properties of tundra soils and nitrogen and carbon pools have been studied for several ecosystems in southern and typical tundra of the Taz Peninsula. Brief description of the selected soil samples is presented in Table 1.

Table 1. List of analyzed soil samples

Sample number	Depth, cm	Brief description of sampling plot	Soil type	
			Russian classification	International classification (WRB)
The southern tundra Plot 1: lower plot with slightly sloping surface, grass-moss tundra				
1	0-10	Cryoturbation spots	Gleezem peaty humic	Gelic Gleysols
2	0-10	Bumps	Gleezem peaty humic	Gelic Gleysols
3	0-10	Plot between bumps	Gleezem peaty	Histi-Gelic Gleysols
Plot 2: elevated (compared with the first) plot with slightly sloping surface, shrub-moss tundra				
4	5-15	Clearings between bushes	Gleezem peaty humic	Gelic Gleysols
5	3-10	Close to bushes	Gleezem peaty	Histi-Gelic Gleysols
Typical tundra Plot 3: the most lowered wet area bumpy mossy tundra				
6	0-10	Bumps	Criozem gleyi	Thixotropic Cryosols
7	0-10	Plot between bumps	Peat-criozem gleyi	Histic Cryosols
Plot 4: lowered (at the foot of the hill) the most humid area of shrub-moss bumpy tundra				
8	0-10	Bumps	Criozem gleyi	Thixotropic Cryosols
9	0-10	Plot between bumps	Peat-criozem gleyi	Histic Cryosols
Plot 5: elevated (near the top of the ridge) dry bumpy area, mossy tundra				
10	0-10	Bumps	Criozem gleyi	Thixotropic Cryosols
11	0-10	Plot between bumps	Peat-criozem gleyi	Histic Cryosols
Plot 6: elevated (on top of the ridge) the most dry plot on sandy ground, lichen-moss tundra				
12	0-10	Meso-elevation	Podbur gleyi	Gleyi-Gelic Cambisols

Selected soil samples were analyzed for several properties characterizing nitrogen in the background soils, as well as indicators that have a significant impact on the processes of transformation of nitrogen compounds in the soil (water pH, the content of physical clay and organic carbon). In southern tundra, the content of total organic nitrogen in the upper, most biologically active part of the soil profile was significantly dependent on modern cryoturbation processes and the resulting position of the soil in microrelief. The soil is fresh (uncovered or poorly covered with vegetation) spots of cryoturbation (sample 1) is characterized by extremely low contents of total nitrogen, which is natural, since the upper part was introduced from the underlying horizons. The soil of adjacent bumps (sample 2) contains 9 times more total nitrogen, and soil of depressions between bumps – (sample 3) is 1.8 times more total nitrogen than the freshly crioturbated soil. It is likely that the content of total nitrogen in the series of soils reflects their modern genesis. The vegetation colonization of the youngest crioturbated soils cryoturbation is accompanied by organic matter accumulation (and with it nitrogen) and thus the bumps with 100% projective cover are formed. Then, after the elevating adjacent parcels as a result of permafrost processes, micro-depressions are formed, where, due to the additional revenues of the dissolved organic matter there is a further accumulation of soil organic matter. Further, depressions are created favorable conditions for the formation of ice lenses, resulting in the substitution of top soil with soil from lower horizons, which has a low content of organic nitrogen (Figure 1).

A similar pattern, although less pronounced, is observed when comparing the number of total nitrogen in the soil for bump and depression plots of the typical tundra.

Mobile soil mineral nitrogen is represented almost exclusively exchangeable ammonium (Table 2). The extremely low content of nitrate nitrogen (less than 4% of the total mineral content of mobile nitrogen) indicates that in soils, both in the southern and typical tundra ecosystem nitrification process is almost completely inhibited. Most likely this is due to the fact that the temperature optimum of this process is 25-35°C.

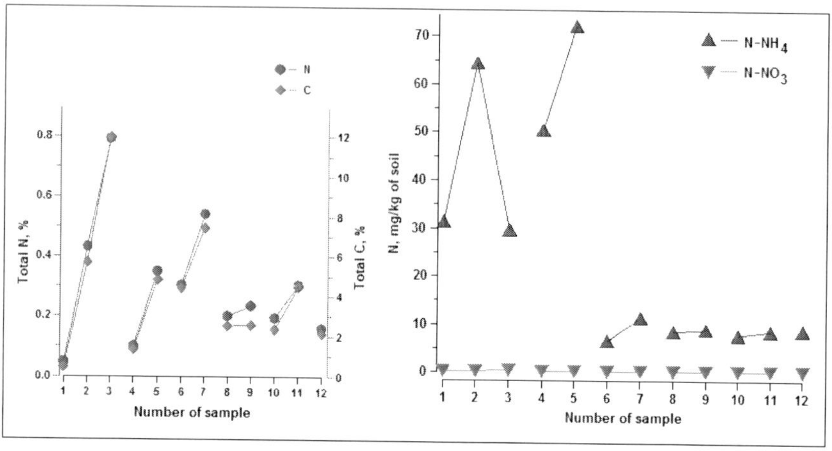

Figure 1. Selected chemical properties of tundra soils (numbers of samples on x-exes are given in Table 1).

The content of exchangeable ammonium in comparison with the total nitrogen to a greater extent was not dependent on the position of the soil on the elements of the microrelief, and most likely was determined by the temperature regime of soils, which is directly related to the intensity of the first stage of mineralization (ammonification) of soil organic matter. It is known that mineralization of nitrogen is severely limited by low temperatures and high levels of soil moisture. In soils of the southern tundra to the time of sampling (the beginning of intensive plant growth) content of the exchangeable ammonium was on average almost 6 times higher than in soils of the typical tundra (where there had been only the beginning of the growing season for most plants).

When comparing the exchangeable ammonium in soils of bumps and depressions of the typical tundra, revealed a trend of increasing content of the nitrogen forms in the depressions. However, in general, the value of the exchangeable ammonium nitrogen was slightly varied around its average content in these soils – 8.7 mg N/kg, which is probably due to the lack of manifestation of the processes of mineralization and absorption of auto- and heterotrophic organisms.

Table 2. Chemical properties of tundra soils

Parameter	Soil sample number					
	1	2	3	4	5	6
Physical clay, %	29.5	27.5	22.1	21.1	16.7	67.2
pH_{H2O}	7.68	5.86	5.95	5.47	5.32	5.72
Total N, %	0.048 *a*	0.433 *cd*	0.793 *e*	0.101 *ab*	0.350 *c*	0.305 *bc*
Organic C, %	0.47 *a*	5.73 *cd*	11.98 e	1.34 *ab*	4.84 *c*	4.43 *bc*
C/N	10 *a*	13 *b*	14 *b*	13 *b*	14 *b*	15 *b*
Exchangeable NH_4, mg/kg	31.4 *b*	64.5 *c*	29.8 *b*	50.6 *bc*	72.2 *c*	6.6 *a*
Exchangeable NH_4, % from total N	6.3 *d*	2.0 *b*	0.4 *a*	4.3 *c*	2.1 *b*	0.2 *a*
$N\text{-}NO_3$, mg/kg	0.3 *b*	0.3 *b*	0.5 *b*	0.3 *b*	0.3 *b*	0.3 *b*

Parameter	Soil sample number					
	7	8	9	10	11	12
Physical clay, %	65.1	29.6	33.5	61.7	62.3	15.3
pH_{H2O}	5.89	6.07	6.17	6.61	6.69	5.81
Total N, %	0.542 *d*	0.202 *b*	0.235 *bc*	0.195 *b*	0.304 *bc*	0.160 *ab*
Organic C, %	7.44 *d*	2.53 *b*	2.55 *b*	2.34 *ab*	4.48 *c*	2.13 *ab*
C/N	14 *b*	12 *ab*	11 *ab*	12 *ab*	15 *b*	13 *b*
Exchangeable NH_4, mg/kg	11.4 *ab*	8.6 *ab*	9.0 *ab*	7.9 *ab*	8.7 *ab*	8.9 *ab*
Exchangeable NH_4, % from total N	0.2 *a*	0.3 *a*	0.3 *a*	0.4 *a*	0.3 *a*	0.6 *ab*
$N\text{-}NO_3$, mg/kg	0.3 *b*	0.2 *a*	0.2 *a*	0.2 *a*	0.3 *b*	0.3 *b*

Note: *a, b, c, d, e* and their combinations – statistical significance of differences between samples.

The content of total nitrogen is highly correlated with total organic carbon (R = 0.99) and nitrate nitrogen (R = 0.82) (Table 3).

Between content of exchangeable ammonium nitrogen and physical clay there was a highly significant negative correlation, which probably indicates the presence of a quite intense non-exchangeable fixation of ammonium by three-layer clay minerals. Significant correlation between the studied forms of nitrogen and the water pH range in studied soils has not been identified.

Table 3. Matrix of pair correlations according to Pearson (6 signs, 12 objects)

	pH_{H2O}	**Total N**	**Organic C**	**Exchangeable NH_4**	**Nitrate N**
Physical clay, %	0.25	0.08	0.07	- 0.58*	- 0.22
pH_{H2O}	-	- 0.33	- 0.32	- 0.34	- 0.07
Total N	-	-	0.99*	0.12	0.82*
Organic C	-	-	-	0.12	0.85*
Exchangeable NH_4	-	-	-	-	0.31

Note * - significant correlation coefficients. Thresholds of equal accuracy at 5% - 0.576, at 1% - 0.708

Thus, characterizing nitrogen regime of tundra soils we can conclude the followings:

1. Mineralization of soil organic matter of both southern and typical tundra ecosystems is limited to the first stage – ammonification. The process of nitrification is almost completely inhibited. The intensity of ammonification at low temperatures in tundra ecosystems is much higher compared to more southern ecosystems under the same temperature regime of soils.
2. Due to the absence of losses of mineral nitrogen due to denitrification, the downward migration of nitrates and volatilization of ammonia, the soil of tundra ecosystems accumulate significant quantities of plant available nitrogen (in the form of exchangeable ammonium), and thus, a very intensive growth and development of plants since the establishment of suitable plant temperature.
3. Plants of the tundra, in contrast to more southern ecosystems, have developed an adaptive mechanism to the predominant absorption of ammonia nitrogen.
4. Content of soil organic matter and associated total nitrogen depends on the position of soil on elements of micro-relief (soil depressions characterized by a higher content of these indicators),

driven by modern genesis of soils as a result of permafrost processes.

5. Content of total nitrogen in soil has a significant positive correlation with organic carbon and nitrate nitrogen. The number of exchangeable ammonium is negatively correlated with the content of physical clay, which probably indicates the presence of a quite intense non-exchange fixation of ammonium by three-layer clay minerals. Significant correlation between the studied forms of nitrogen and water in the pH range studied soils has not been identified.

Although the obtained values of total carbon and nitrogen in soil were generally within the range of values reported for the Arctic ecosystems (Manakov 1972; Post & Emanuel 1982; Lipson et al. 1999; Bashkin 2014), nitrogen availability, microbial biomass nitrogen and mineralization rate of total nitrogen were clearly unevenly distributed between the different forms of the microrelief in bumpy tundra and along the latitudinal gradient of studied plots (southern and typical tundra). In large scale, soils of depression zones in nearly all areas show higher pools of nitrogen and rate of nitrogen turnover, compared with bumps. We emphasize that our findings are the first report on clear differences in the dynamics of nitrogen between the bumps and adjacent depression areas. However, for similar ecosystems – tussock tundra, earlier it was reported that higher concentrations of total nitrogen and mineral nutrition elements were found in the areas between hummocks, in comparison with soils of hummocks (Johnson et al. 1996). Recall that the tussocks are small bumps, which formed by cottongrass, sedge and are the most comparable type of surface with bumps.

Two factors may explain the differences in the dynamics of nutrients between microenvironments of the bumpy tundra. First, the microrelief and associated physico-chemical properties of the landscape (such as soil moisture, temperature, pH) can determine the nature of the (re)distribution and mineralization of nitrogen. Secondly, vegetation can variously affect the regime of nutrients in soils of bumps and depression areas.

To study the effect of meso- and microrelief affecting the supply of moisture in the soil, we conducted a study of major forms of nitrogen pools (nitrate, exchangeable ammonium, nitrogen of soil microorganisms biomass, nitrogen of amino acids and non-biotic organic nitrogen) in soils of the southern and typical tundra (Figure 2).

Emphasis was placed on comparative studies of nitrogen fund depending on latitudinal gradient (and consequently the average temperature), meso-relief (have a big influence on soil moisture) and surface. We emphasize that one of the characteristics of tundra ecosystems is the heterogeneity of the microrelief, which is caused by the presence of bumps and depressions between bumps formed as a result of various cryogenic processes. Effects of microrelief and the closely related differentiation of vegetation over-cover effects of temperature (during the transition from the southern tundra to typical one) and soil moisture (or the effect of meso-relief) and acts as the main controlling factor of the nitrogen fund in tundra soils.

Further details will focus on the effects of microrelief and soil conditions. Litter of the Arctic plants is usually redistributed from increases to the lower parts of the landscape, defining, thus, a pathway for important nutrients in a certain area (Fahnestock et al. 2000). On studied plots, the litter of most plant species was accumulated in areas between bumps, providing organic matter to the soil. Nitrogen turnover was higher in soils between bumps compared to the bumps that prove a higher rate of mineralization of nitrogen. Therefore, higher amounts of total nitrogen and carbon, discovered in areas between bumps, due to the increasingly large external supply of substrates than lower rates of decomposition. The relief further supports lateral flow of water from bumps to depression area (Quinton et al. 2000) and, thus, the accumulation of dissolved organic and inorganic compounds in the depressions between the bumps. Both factors, the lateral inflow of nitrogen and higher speed of nitrogen cycling, probably contribute to greater availability of mineral and organic pools of nitrogen in zones between bumps.

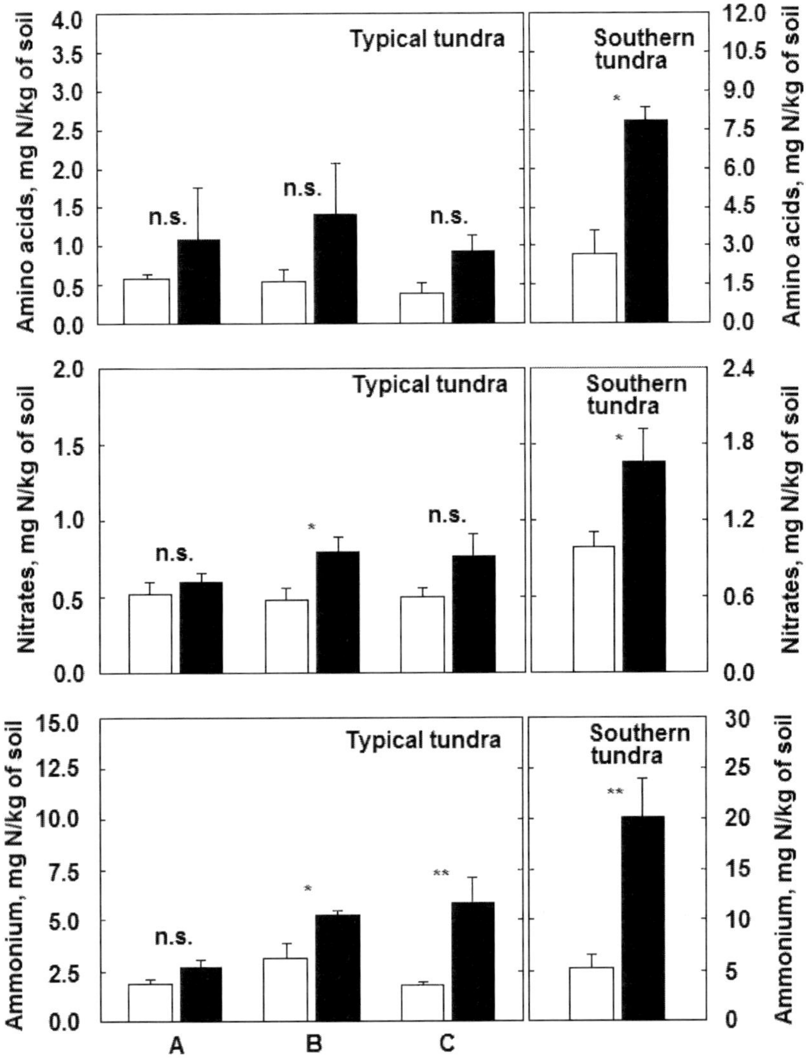

Figure 2. Content of mobile forms of nitrogen in tundra soils: A – waterlogged plot, B – wet plot, C – moderately wet plot (* $p < 0.05$, ** $p < 0.01$, n.s. – not significant).

Some physicochemical parameters (such as temperature and soil moisture, pH) also vary with micro-topography. It is known that temperature is one of the main factors influencing the activity of transformation processes carried out by microorganisms. In our case, the overall rate of mineralization of nitrogen was, on average, 3 times higher in soils placed in areas between bumps compared to soil mounds. These differences were also reflected in the values of microbial biomass, which values are significantly positively correlated ($r^2 = 0.54$, $p<0.005$) with a rate of total mineralization of nitrogen. Similar patterns in distribution of microbial biomass were obtained for the tussock tundra with a higher content of microbial biomass in the soil between tussocks, in comparison to the soils of the bumps. Paradoxically, zones between bumps are more humid, and therefore cooler habitats compared with the bumps (Zoltai & Tarnocai 1974). The colder temperature in the soil placed in zones between bumps is confirmed by different depth of unfrozen soil layer, which is significantly higher in bumps compared with areas between them. It is known that low temperatures inhibit microbial activity and consequently the release of nutrients into the soil (Lloyd et al. 1994). In addition, in very wet conditions or in case of flooding, characteristic for zones between bumps, microbial activity is usually limited (Johnson et al. 1996). So we came to the conclusion that soil temperature and moisture content may be responsible for the observed differences in mode of nutritional elements in the bumps and areas between bumps. However, regional scale, nitrogen availability and total mineralization was significantly higher for the area of the southern tundra, compared to the typical one. In this regard, there is the assumption of indirect (or implicit) effect of temperature on soil processes. Probably higher productivity of these ecosystems has a positive effect on the pools of soil nitrogen, microbial biomass and the rate of mineralization of nitrogen, which are also highly correlated in our studies (Figures 3 and 4).

The obtained values of the total mineralization rate for two the Arctic ecosystems was quite high, especially in the southern tundra, where they reached 20 mg N/kg soil per day (in the typical tundra – 6 mg N/kg soil per day). These values are comparable to the velocities measured in soils of tropical rain forests and boreal forests: 27 and 16 mg N/kg soil per day, respectively (Fisk & Fahey 2001). In contrast, such types of non-forested ecosystems, like boreal and the Alpine meadows showed a much smaller average values of mineralization rates: 5 and 8 mg N/kg soil per day, respectively (Corre et al. 2002; Shaw & Harte 2001). Data on the rate of mineralization in the Arctic regions are extremely rare, but are reported to lie in the upper range of average values for all terrestrial ecosystems in the world data on nitrogen total mineralization rates – 13 mg N/kg soil per day (Bashkin et al. 2012). Only recently, it was proposed that the decomposition of organic matter is directly (explicitly) associated with average annual temperature only at the global (smaller) scale.

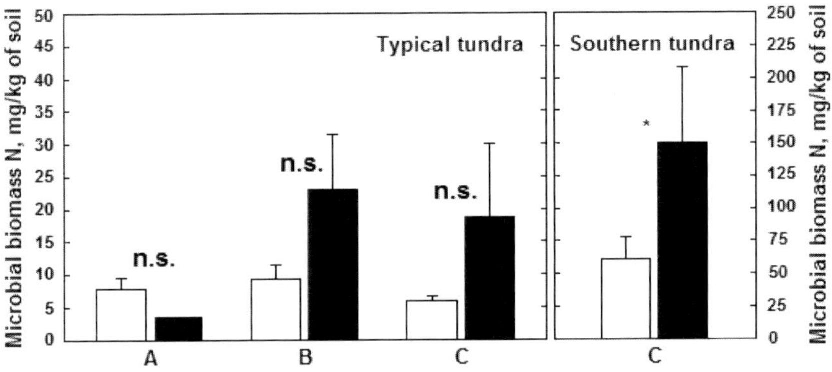

Figure 3. Nitrogen content of microbial biomass in tundra soils: A – waterlogged plot, B – wet plot, C – moderately wet plot (* $p < 0.05$, ** $p < 0.01$, n.s. – not significant).

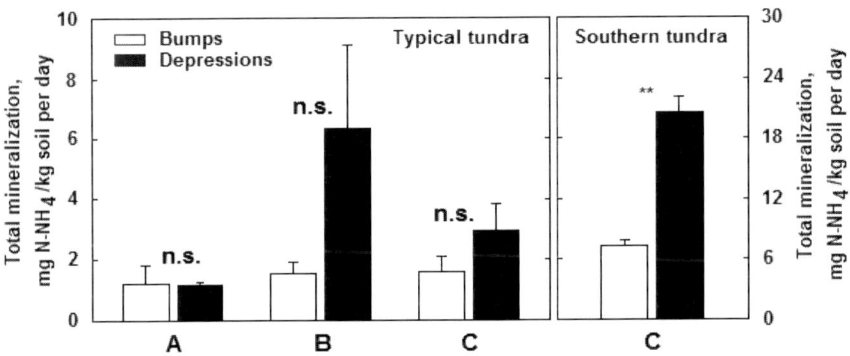

Figure 4. Total mineralization rates in tundra soils: A – waterlogged plot, B – wet plot, C – moderately wet plot (* $p < 0.05$, ** $p < 0.01$, n.s. – not significant).

Most likely, this is due to the fact that microbial activity is primarily controlled by other factors such as substrate availability and quality (Giardina & Ryan 2000). Similarly, in our studies, the observed differences in the pools of nitrogen and the speed of transformation between the bumps and depression areas did not correlate with the temperature and humidity of the soil and, therefore, were due to differences in position on elements of the microrelief, which leads to a redistribution of litter and flux of dissolved nutrients.

As already noted above, the content of soil moisture could not explain the differences in the concentration of nutrients and speeds of their cycling in various micro-relief sites of bumpy tundra. However, for depression soils of typical tundra when compared to waterlogged, wet and moderately wet plots the resulting trend of reduced availability of nitrogen and the speed of the cycle with increasing humidity. This indicates the inhibitory effect of flooding on soil microbial activity. A similar trend cannot be observed in the bumps because their water regime in tubercles varies slightly in waterlogged, wet and moderately wet soils/habitats due to elevated positions in the microrelief.

Although it can be assumed that the speed of the nitrogen cycle can strongly depend on the pH of the soil this explanation can be ruled out for the observable differences in the dynamics of nitrogen, as significant differences in soil pH in bumps and depressions between bumps were not observed.

Thus, the processes of microbiological transformation of nitrogen compounds and the availability of nutrients were higher in areas between bumps, compared with bumps, largely due to increased reserves of the substrate than soil temperature, moisture, or pH. It is also consistent with other studies, where higher microbial activity was obtained in soils with a high litter accumulation and higher content of available dissolved organic and mineral nutrients (Wang & Bakken 1997; Fahnestock et al. 2000; Bashkin et al. 2012).

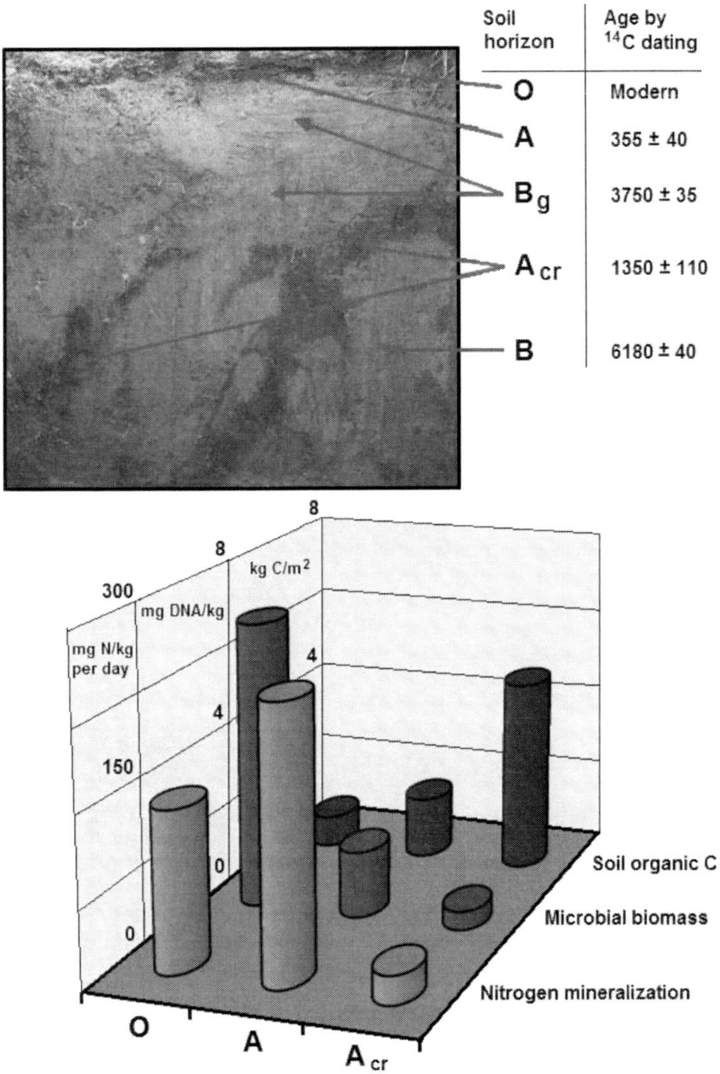

Figure 5. Stock of soil organic carbon, microbial biomass and mineralization of nitrogen in different horizons of tundra soil: O – litter, A and A_{cr} – modern and buried horizons of accumulation of soil organic matter).

In addition to the heterogeneous distribution of nitrogen pools in the plane (i.e., on the surface of soils) mosaic of stocks of different pools of nitrogen was observed in the soil profile with depth due to constantly occurring in tundra soils cryoturbation processes. In the photograph (Figure 5), as an example, presents the profile of tundra soil with displaced organic-accumulative horizon A by cryogenesis at different depths. Despite the high content of total nitrogen in organo-accumulative fragments of crioturbated horizon, the rate of mineralization of nitrogen was on average 7.5 times higher in the horizon A, compared with the fragments of this horizon in the lower part of the soil profile. This difference may be due to differences in temperature to the actual horizon and its fragments, located significantly closer to the layer of permafrost and, consequently, receiving much less heat.

CONCLUSION

Thus, the work performed has allowed to establishing the regularities of the heterogeneous distribution of reserves of different forms of nitrogen in soils of tundra ecosystems in space – depending on the meso- and microrelief, and with increasing depth in the soil profile. At regional (small) scale, the temperature, of course, controls the dynamics of nitrogen in the Arctic ecosystems, which is proved by the higher mean values of the pools of nitrogen and the speed of the cycle in southern tundra compared to the typical one.

Soil moisture (or the effect of meso-relief) affects the mode of nitrogen in only the soil between bumps along the gradient from temperate-humid to extra-humid conditions. However, in a larger scale the effect of microrelief and the closely related differentiation of vegetation over-cover the effects of temperature and soil moisture. Therefore microrelief acts as the main controlling factor of the nitrogen fund in tundra soils due to the redistribution of litter and the lateral input of dissolved organic and mineral substrates at depression zone. The mosaic distribution of soil nitrogen pool and its spatial mineralization rate is enhanced by cryoturbation processes that result in periodical leaching of fragments of organo-accumulative horizon with high nitrogen content deeper into the soil. This is processing in conditions of lower temperatures and, accordingly, reduced activity of the microbiota.

ACKNOWLEDGMENT

The authors express their deep appreciation and gratitude to Dr. Anatoly K. Arabsky (Gazprom Dobycha Yamburg LLC) for the financial support of this research.

CONFLICT OF INTEREST

The authors confirm that this article content has no conflicts of interest.

REFERENCES

Bashkin, VN (1987). *Nitrogen Agrogeochemstry*. ONTI Publishing House, Pushchino.

Bashkin, VN (2002). *Modern Biogeochemistry*. Kluwer Academic Publishers, Dordrech-London-Boston.

Bashkin, VN (2006). *Modern Biogeochemistry: environmental risk assessment*, 2nd edition. Springer Publishers, the Netherlands.

Bashkin, VN (2014). *Biogeochemistry of Polar Ecosystems in Gas Industry Impacted Zones*. Gazprom VNIIGAZ Publishing House, Moscow.

Bashkin, VN (2016). *Biogeochemical Technologies for Managing Pollution in Polar Ecosystems*. Springer Publishers, Switzerland.

Bashkin, VN, Arno, OB, Arabsky, AK, Barsukov, PA, Priputina, IV, & Galiulin, RV (2012). *Retrospective and Forecast of Geoecological Situation on the Gas Condensate Fields of the Far North*. Gazprom VNIIGAZ Publishing House, Moscow.

Fahnestock, J, & Romanovsky, VE (2005). Winter biological processes could help convert arctic tundra to shrubland. *Bioscience*, **55**, 17-26.

Fisk, MC, & Fahey, TJ (2001). Microbial biomass and nitrogen cycling responses to fertilization and litter removal in young northern hardwood forests. *Biogeochemistry*, 53, 201–223.

Giardina, CP, & Ryan, MG (2000). Evidence that decomposition rates of organic carbon in mineral soil do not vary with temperature. *Nature*, 404, 858-861.

Johnson, LC, Shaver, GR, Giblin, AE, Nadelhoffer, K., Rastetter, ER, Laundre, JA, & Murry, GL (1996). Effects of drainage and temperature on carbon balance of tussock tundra microcosms. *Oecologia*, 108, 737-748.

Le Dize, S, Kwiatkowski, BL, Rastetter, EB, Hope, A, Hobbie, JE, Stow, D, & Daeschner, S (2003). Modelling biogeochemical responses of tundra ecosystems to temporal sand spatial variations in climate in the Kuparuk River Basin (Alaska). *J. Geophys. Res.*, 108(D2), doi:10.1029/2001JD000960.

Lipson, DA, Schmidt, SK, & Monson, RK (1999). Links between microbial population dynamics and nitrogen availability in an alpine ecosystem. *Ecology*, 80, 1623-1631.

Lloyd, J, & Taylor, JA (1994). On the temperature dependence of soil respiration. *Functional Ecology*, 8, 315-323.

Manakov, KN (1972). *The Biological Cycle in Arctic and Tundra Ecosystems*. Nauka Publishing House, Moscow.

Moskovchenko, DV (2010). *Geochemistry of Landscapes of the North of the West Siberian Plain: Structural-Functional Organization of the Substance of Geosystems and Problems of Ecodiagnostic*. The dissertation on competition of a scientific degree of doctor of geographical sciences. SPU Publishing House, St-Petersburg.

Oechel, WC, Vourlitis, GL, Hastings, SJ, Zulueta, RC, Hinzman, L, & Kane, D (2000). Acclimation of ecosystem CO_2 exchange in the Alaskan Arctic in response to decadal climate warming. *Nature*, 406, 978-981.

Post, WM, & Emanuel, WR (1982). Soil carbon pools and world life zones. *Nature*, 298, 156-9.

Quinton, WL, Gray, DM, & Marsh, P (2000). Subsurface drainage from hummock-cover hillslopes in the Arctic tundra. *Journal of Hydrology*, 237, 113–125.

Shaw, RM, & Harte, J (2001). Response of nitrogen cycling to simulated climate change: differential responses along a subalpine ecotone. *Global Change Biology*, 7, 193-210.

Wang, J, & Bakken, LR (1997). Competition for nitrogen during decomposition of plant residues in soil: effect of spatial placement of N-rich and N-poor residues. *Soil Biology and Biochemistry*, 29, 399-409.

Zoltai, SC, & Tarnocai, C (1974). Soil and vegetation of hummocky terrain. *Environmental-Social Committee, Task Force on Northern Oil Pipeline Development, Report*, 74-3.

In: Ecological and Biogeochemical Cycling … ISBN: 978-1-53612-081-3
Editor: Vladimir N. Bashkin

Chapter 3

ASSESSING THE *IN SITU* EMISSION OF GASES IN TUNDRA ECOSYSTEMS

***Vladimir N. Bashkin*[1,2], *Rauf V. Galiulin*[3] and *Pavel A. Barsukov*[4]**

[1]Institute of Physicochemical and Biological Problems of Soil Science of Russian Academy of Sciences, Pushchino, Russian Federation
[2]Gazprom VNIIGAZ LLC, Razvilka, Russian Federation
[3]Institute of Basic Biological Problems of Russian Academy of Sciences, Pushchino, Russian Federation
[4]Institute of Soil Science and Agrochemistry of Siberian Branch of Russian Academy of Sciences, Novosibirsk, Russian Federation

ABSTRACT

This chapter is devoted to assessing the *in situ* emission of gases in tundra ecosystems using outdoor chambers with experimental elevated temperature and control plots. The rate of CO_2 emissions is greatly increased due to warming and leads to a net loss of carbon from the "soil-plant" system at elevated temperatures compared with normal temperatures. The increase of plant biomass to warming was largely due to the response of lichens as the dominant component of the dwarf shrub-lichen tundra.

Keywords: CO_2 emissions, outdoor chambers, tundra ecosystems, *in situ* monitoring

INTRODUCTION

Considering global climate change and its impact on the Arctic ecosystems, it has been hypothesized that warming will lead to significant vegetation change: to reduce the biomass of mosses (Chapin et al. 1995b) and a significant increase in the proliferation of shrubs (Sturm et al. 2001). In addition, warming can cause changes in the intensity cryoturbation processes, giving, thus, the change in the result of transformation processes in the Arctic soils (Vasilevskaya 1996). Taken together, these changes in the environment can significantly affect the structure and function of bumpy tundra, but the magnitude and sequence of such changes is still unclear, since our knowledge of the patchiness of this complex system is very limited. Because the Arctic ecosystems store large stocks of soil carbon (Post & Emanuel 1982), changes in biogeochemical cycles can also be very important when considering a reverse correlation with climate warming. Detailed studies of the nitrogen of the background soils of tundra ecosystems and its heterogeneity in space are the necessary basis to study the effects of elevated temperatures on biogeochemical cycles of nitrogen and carbon.

Earlier field monitoring data in gas industry impacted zones of the Taz Peninsula did not show any anthropogenic influence on methane and carbon dioxide emissions (Bashkin et al. 2012; Bashkin 2014; Bashkin & Barsukov 2016).

With global warming, it is assumed that the decomposition of soil organic matter (SOM) will be accelerated and therefore will increase the release of carbon from soils to the atmosphere (Oechel et al. 1993). Enriched CO_2 atmosphere may increase global warming due to feedback effects, despite an equivalent increase of production net primary production, offsetting the accumulation of CO_2. However, it remains unclear what amount of carbon can potentially be released from soil due to global warming. Furthermore, in the north of West Siberia at a regional scale, the opposite process of climate cooling is observed (Bashkin & Galiulin 2016) and this should be taken into account as well.

METHODS

In situ study of the effect of elevated temperatures in the processes of the nitrogen cycle and the carbon and productivity of tundra plants was performed using the open hexagonal chambers according to the standard ITEX (International Tundra Experiment). Hexagonal chambers were installed in tundra ecosystems on the Taz Peninsula approximately 5 km from Yamburg gas production point. It is known that in these cameras the average temperature of surface air is increased by an average of 2-3°C in comparison with the adjacent territory, allowing in natural field conditions to investigate the effects of elevated temperatures (Bashkin et al. 2012).

FIELD MONITORING RESULTS

The dynamics of the rate of emission of carbon dioxide from control plots and open chambers were studied during the day. In the presented results the carbon dioxide emissions were due to the respiration of plants and soil (soil microorganisms), less absorption of CO_2 during photosynthesis (data on emission of the inhibition of photosynthetic activity are currently being processed). Figure 1 presents data on average for 15 days.

In general, the rate of emission is directly dependent on temperature due to incoming solar radiation. However, in case of control plots, the dynamics of CO_2 emissions was expressed quite mild during the day (varying within a narrow range from 0.02 to 0.04 μm/m^2 per hour), the outdoor cameras emissions rates were changed on average from 0.02 to 0.19 μm/m^2 per hour. The maximum emission was observed at around 3 pm. By 8 pm, the rate of emission is significantly reduced to 0.06-0.07 μm/m^2 per hour, but still remained higher compared to control. In the evening and night hours (until 3 am) the rate of CO_2 emissions remained approximately at the same level. From 3 am to 9 am there was a gradual decrease in the rate of emission to values typical for the control plots. After that, there was a sharp increase in the rate of emission with a maximum at 3 pm.

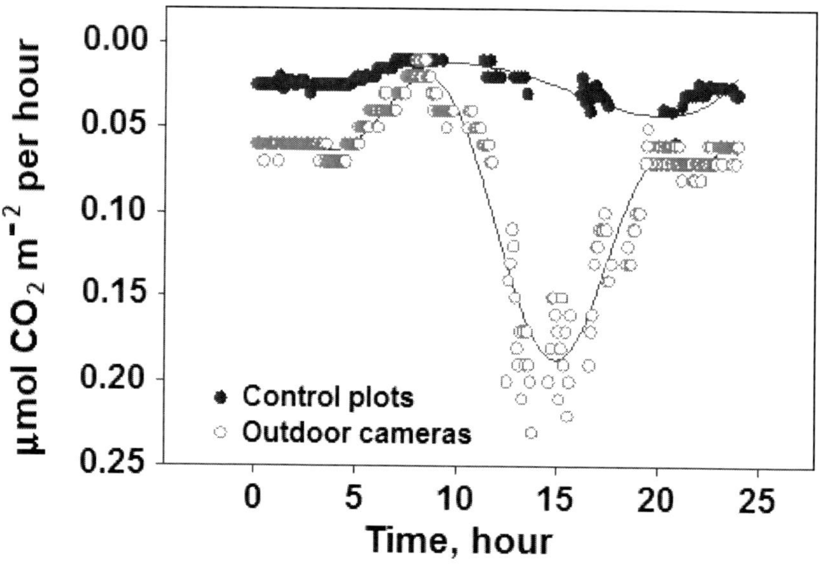

Figure 1. Dynamics of the rate of CO_2 emissions from control plots and outdoor cameras during the day (average data for 15 days).

Calculated over the observation period the total emission of carbon dioxide allows to making a preliminary conclusion: despite a small increase in average temperature (by 2-3 degrees), rate of carbon dioxide emissions in an average day increased 6-fold compared to control, reaching a value of almost 1 g CO_2/m^2/day (Figure 2).

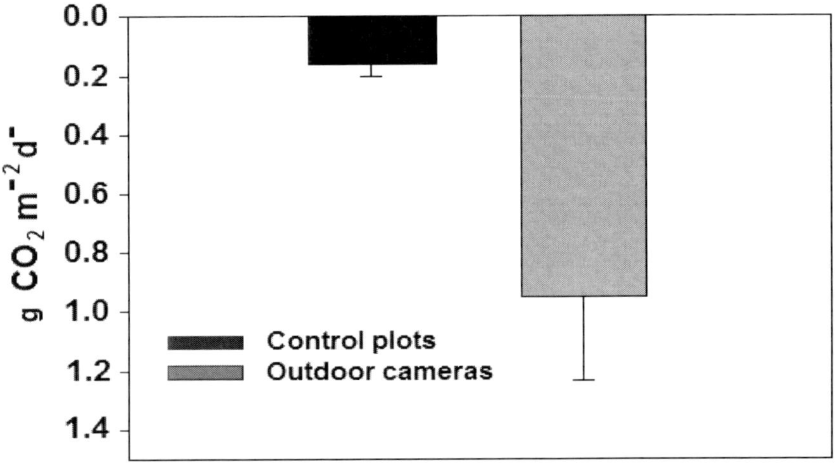

Figure 2. Total for the day CO_2 emissions from control plots and outdoor cameras (average data for 15 days in 3 field replicates).

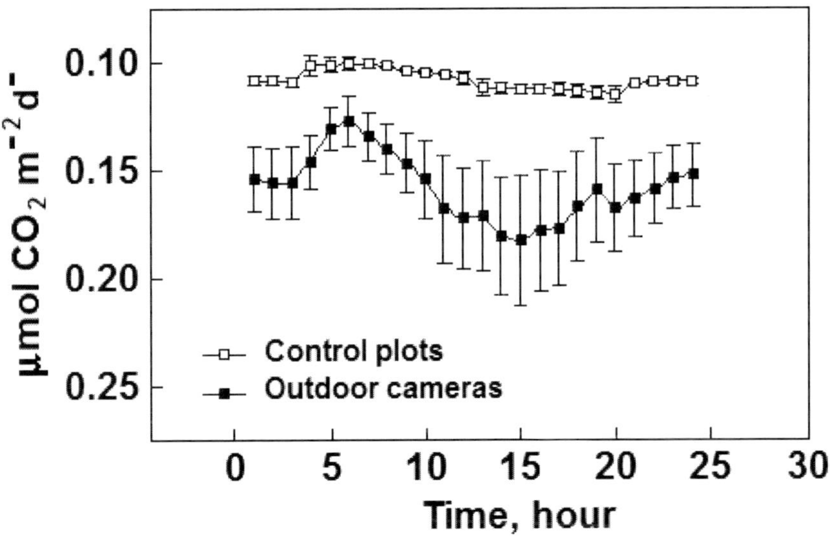

Figure 3. Changing CO_2 balance in ecosystem during the day after 2 years of the experimental warming.

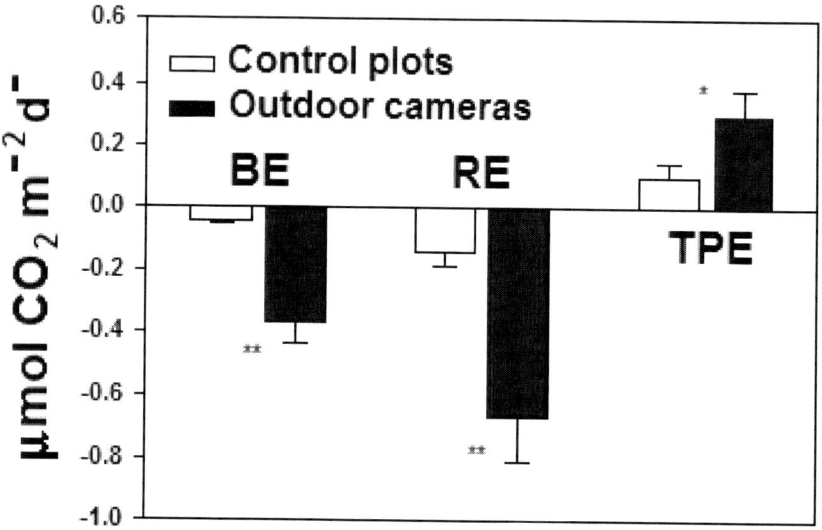

Figure 4. Effect of experimental warming on the CO_2 balance of the ecosystem (BE) at maximum illumination, total "respiration of the ecosystem" (RE) and total photosynthesis of the ecosystem (TPE) in the dwarf shrub-lichen tundra (* -p < 0.05, ** -p < 0.01).

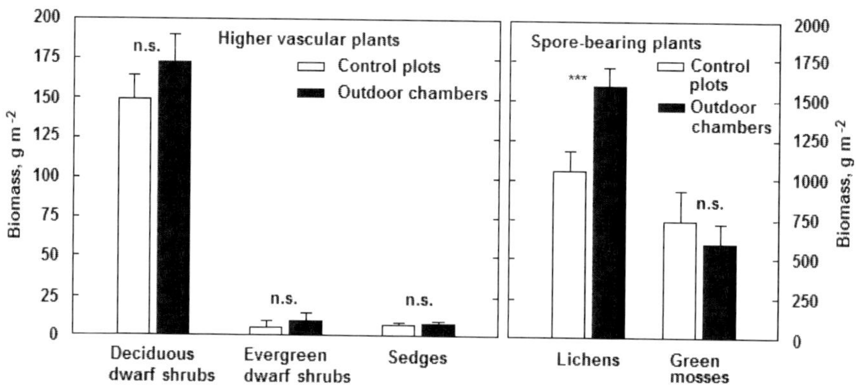

Figure 5. Effect of experimental warming on aboveground biomass of plants of the dwarf shrub-lichen tundra (*** p < 0.005; n.s. - not certain).

The carbon content in the horizon of accumulation of soil organic matter significantly decreased by 4.7% in outdoor cameras, compared with the control, while the nitrogen content increased and overall warming effect on the availability of soil nitrogen was significant.

The trend of the CO_2 flux during the day was similar for outdoor cameras and control plots (Figure 3).

Net CO_2 balance in the ecosystem (i.e., the difference between its sink and emission) was positive early in the morning along with the beginning of photosynthesis, which means that the sink of CO_2 into "soil-plants" system.

During the day, ecosystem breathing dominated by photosynthesis showing negative values of CO_2 exchange rate. Negative values on the graph respectively represent the net emission of CO_2 into the atmosphere. At control plots the daily CO_2 balance however was close to zero, i.e., the rate of photosynthesis and respiration were balanced in the study period. The average values of net CO_2 balance at the maximum illumination were significantly lower in the outdoor chambers than in the control, which means increased CO_2 emissions to the atmosphere from these chambers (Figure 4). Total carbon losses from "soil-plant" system were 8 times higher in the outdoor chambers compared with the control plots. Accordingly, the respiration: of the ecosystem in outdoor chambers increased by 5 times, and the rate of photosynthesis - 3 times.

As already noted, the aboveground plant biomass significantly increased after two years of experimental warming.

However, related to increased photosynthetic fixation of carbon is not compensated for the much greater losses of CO_2 from ecosystems due to experimental warming that may be caused by the increased respiration to both plants and heterotrophic soil microbiota. The C/N ratio in the plant biomass was unchanged. Plant growth, therefore, was not limited by nitrogen at elevated temperatures, but other nutrients such as phosphorus might limit growth of tundra plants.

The impact of experimental warming on the biomass of individual species is shown in Figure 5.

Aboveground plant biomass significantly increased after two years of experimental warming, mainly due to the lichens that stood out for their strong response to temperature increases and accounted for 95% of the total aboveground plant biomass. Deciduous dwarf shrubs (*Betula nana, Vaccinium uligunosum, Salix glauca*), an evergreen dwarf shrubs (*V. vitis idaea, Ledum decumbens*), sedges and mosses did not show significant changes in biomass with warming. However, the leaf surface of *Betula nana* significantly increased by 63% in the outdoors chambers compared to control plots. The total content of C and N in aboveground plant biomass was also significantly higher in outdoors chambers compared with the control plots.

CONCLUSION

Thus, the increase of plant biomass for a warming effect is largely due to the response of lichens, and this is the dominant component of the dwarf shrub-lichen tundra. The rate of CO_2 emissions is greatly increased due to warming and leads to a net loss of carbon from the "soil-plant" system at elevated temperatures compared with normal temperatures. This confirms the hypothesis that, as a result of climate warming, the net carbon balance of the Arctic ecosystems (at least in the short term) could change from a net sink of carbon to its clean emissions. The responses of individual processes to the warming likely depend on the type of the system under investigation, which requires experimentation with the manipulation of temperature in a variety of plant communities and soil types.

With this in mind, the change of temperature processes on the observed cold is able to restore the status quo in the biogeochemical cycles of carbon and nitrogen, including CO_2 emission and sink.

CONFLICT OF INTEREST

The authors confirm that this article content has no conflicts of interest.

ACKNOWLEDGMENT

The authors express their deep appreciation and gratitude to Dr. Anatoly K. Arabsky (Gazprom Dobycha Yamburg LLC) for the financial support of this research.

REFERENCES

Bashkin, VN (2014). *Biogeochemistry of Polar Ecosystems in Gas Industry Impacted Zones*. Gazprom VNIIGAZ Publishing House, Moscow.

Bashkin, VN (2016). *Biogeochemical Technologies for Managing Pollution in Polar Ecosystems*. Springer Publishers, Switzerland.

Bashkin, VN, Arno, OB, Arabsky, AK, Barsukov, PA, Priputina, IV, & Galiulin, RV (2012). *Retrospective and Forecast of Geoecological Situation on the Gas Condensate Fields of the Far North*. Gazprom VNIIGAZ Publishing House, Moscow.

Bashkin, VN, & Barsukov PA (2016). *Emission of carbon dioxide and methane in gas industry impacted ecosystems*. In: Bashkin, VN (Ed). Biogeochemical Technologies for Managing Pollution in Polar Ecosystems. Springer Publishers, Switzerland, 65-72.

Bashkin, VN, & Galiulin, RV (2016). *Climate cycling and modelling in polar ecosystems*. In: Bashkin, VN (Ed). Biogeochemical Technologies for Managing Pollution in Polar Ecosystems. Springer Publishers, Switzerland, 95-108.

Chapin, FSI, Shaver, GR, Giblin, AE, Nadelhoffer, KJ, & Laundre, JA (1995b). Responses of arctic tundra to experimental and observed changes in climate. *Ecology*, 76, 694-711.

Oechel, WC, Hastings, SJ, Vourlitis, GL, Jenkins, M, Riechers, G, & Grulke, N (1993). Recent change of arctic tundra ecosystems from a net carbon dioxide sink to a source. *Nature*, 361, 520-3.

Post, WM, & Emanuel, WR (1982). Soil carbon pools and world life zones. *Nature*, 298, 156-9.

Sturm, M, Racine, C, &Tape, K (2001). Climate change - increasing shrub abundance in the Arctic. *Nature*, 411, 546-7.

Vasilevskaya, VD (1996). Assessment of the sustainability of tundra permafrost-affected soils to anthropogenic influences. *Bulletin of Moscow University, Ser. 17, Soil Science*, 1, 27-35.

In: Ecological and Biogeochemical Cycling … ISBN: 978-1-53612-081-3
Editor: Vladimir N. Bashkin

Chapter 4

ECOLOGICAL INTERACTIONS IN BIOGEOCHEMICAL CYCLES OF CARBON AND NITROGEN

Vladimir N. Bashkin[1,2], Rauf V. Galiulin[3] and Pavel A. Barsukov[4]

[1]Institute of Physicochemical and Biological Problems of Soil Science of Russian Academy of Sciences, Pushchino, Russian Federation
[2]Gazprom VNIIGAZ LLC, Razvilka, Russia Federation
[3]Institute of Basic Biological Problems of Russian Academy of Sciences, Pushchino, Russian Federation
[4]Institute of Soil Science and Agrochemistry of Siberian Branch of Russian Academy of Sciences, Novosibirsk, Russian Federation

ABSTRACT

The ecological interactions in biogeochemical cycles of carbon and nitrogen were the aim of this paper. The evidence that a stable soil organic matter is preferably decomposed at high temperatures can be extremely important for predicting the impact of climate change on the carbon cycle in the Northern ecosystems. Two points are the most important: a) even the most stable fractions of the soil organic matter can be potentially mineralized by soil microbes in tundra ecosystems; and b) resistant fractions of the soil organic matter can be mineralized in a relatively high degree. Furthermore, the mineralization of resistant carbon compounds can be explained by changes in the structure of microbial communities dominated by microorganisms with high temperature optimum, which is capable of utilizing more sustainable carbon substrate. The paper results prove that temperature, not the depletion of the pool of readily available carbon compounds lead to a preferred degradation of stable fractions of the soil organic matter. Thus, the inventory of stable carbon in tundra biome may not be permanent. These soil organic matter values may grow during the cyclic periods of regional temperature increase and may decline in the relevant cold periods.

Keywords: biogeochemical cycles of carbon and nitrogen, soil organic matter transformation, tundra ecosystem, ecological interactions

INTRODUCTION

Since temperature is one of the key limiting factors for all abiotic and biotic processes, climate change in the Arctic ecosystems is of great importance for global biogeochemical processes. Increased temperature can affect tundra plants directly by reducing the time of preservation of the snow cover and increase in vegetation period and indirectly by changing other factors such as availability of nutrients and soil moisture. On the contrary, low soil temperatures in high latitudes and high humidity impede the processes of mineralization in the soil and therefore are the cause of the limitations of nutrient resources for growth of the Arctic plants. The increase in the rate of mineralization of soil organic matter due to the temperature increase may thus increase the pool of nutrients available to tundra plants that will continue to have a positive effect on their growth and productivity.

As already noted above (Bashkin & Galiulin 2017, in this thematic issue) the functioning of the Arctic ecosystems is largely dependent on the speed of carbon and nitrogen cycling processes. In general, the Arctic ecosystems are characterized by low pool of available nutrients due to the cold and wet conditions and, thus, a low amount of net primary production (McKane et al. 1997; Wookey 2002). Considering global climate change and its impact on the Arctic ecosystems, it has been hypothesized that warming will lead to a significant change in vegetation cover: decline in biomass of mosses (Chapin et al. 1995a) and a significant spread of shrubs (Sturm et al. 2001). In turn, this should lead to a change in the nitrogen cycle in the Arctic ecosystems (and therefore in the functioning of these ecosystems), though, because the speed of assimilation of different nitrogen forms by specific functional groups of plants, must be significantly different. For the prediction of these changes the understanding of nitrogen and carbon cycling in soils and the major plants of the Arctic ecosystems is important.

Relationship between Climate Change and Biogeochemical Cycles of Nitrogen and Carbon

With global warming it is assumed that the decomposition of soil organic matter (SOM) will be accelerated and therefore will increase the release of carbon from soils to the atmosphere (Oechel et al. 1993). Enriched CO_2 atmosphere may increase global warming due to feedback effects (greenhouse effect). However, it remains unclear what amount of carbon can potentially be released from the soil. In addition, the question remains whether enhanced mineralization of more stable carbon source at elevated temperature by itself, or is it due to the depletion of readily available substrates. Also for understanding the mechanisms underlying the utilization of different carbon sources is important to study the changes in the structure of microbial communities. In the contrary, temperature decreasing will lead to SOM transformation processes at steady state conditions.

Although microbial activity and thus decomposition processes are known to be temperature-dependent, the effect of warming on soil processes remains controversial. As observed, the rate of decomposition of

SOM remains largely constant with increasing temperature. It is most likely that *in situ* the quantity and quality of substrate is more important for speed control of mineralization than temperature (Giardina & Ryan 2000). It was suggested that only the decomposition of a small labile SOM fraction by heterotrophs is dependent on temperature, while the mineralization is older and more resistant to decomposition fractions of the SOM (proportion which is substantially more than labile SOM) is tolerant to temperature (Liski et al. 1999). Thus, with continued warming, when readily available carbon pools to dwindle, the overall rate of decomposition may decrease (Luo et al. 2001; Rustad 2001; Bashkin et al. 2012).

Since a large proportion of the SOM is composed of more resistant material, sensitivity to temperature and potential availability of this pool of soil carbon is indeed of great importance in the study of climate change. In accordance with the Parton (1987) SOM is divided into 3 large components: (1) small (around 5%) labile pool SOM, which contains microbial biomass and easily decomposable components; (2) large intermediate pool (60-85%), which includes the more stable components of the SOM; and (3) smaller passive pool (10-40%) which consists mainly of resistant organic components that are physically and chemically protected. Because the passive pool is the most inert to biological attack, approximately 20-25% of heterotrophic respiration, likely to arise from the decomposition of the intermediate pool and the remaining 75-80% of the decomposition of the active pool (Townsend 1997; Semenov & Cobut 2015).

But a low degradability and a low sensitivity to temperature of the passive pool reinforces the hypothesis that biogeochemical models which consider the SOM as a homogeneous pool, vastly overestimated the impact of global warming on the processes of biophilic elements circulation (Gu et al. 2004). However, data that support this hypothesis are mostly not experimental, being based on modeled and indirect methods of determination. Direct analysis of the pools of carbon utilized by the microorganisms at different temperatures, is extremely rare and often gives opposite results. For example, Weintraub (2003) used a chemical fractionation of the Arctic soils for showing that different fractions of the SOM are basically more or less equally accessible to soil microorganisms. There is also some empirical evidence that the old sustained SOM is even more sensitive to temperature changes than labile young matter.

The most reliable means of determining the substrates, which are utilized by microbes, is the isotopic analysis of breath CO_2. This is perhaps due to the fact that the ratio of isotopes in the exhaled CO_2 by soil decomposing microbes corresponds to that in the SOM. Despite the obvious benefits of this approach there are only few published data on this topic, most of which show that old and stable pools of organic carbon mineralized more rapidly at higher temperatures (Andrews et al. 2000; Bol 2003). In accordance with this decomposition of the old passive SOM pool is not resistant to temperature change. However, the response appears to be specific for different soils and datasets are often incomplete and need to be cautious when interpreting the results of isotopic analysis due to possible fractionation of the SOM. In addition, the question remains whether the enhanced utilization of more sustainable sources of carbon at high temperatures by itself, or it is a consequence of depletion of readily available substrates and probably, thus, can be inverted when reaching the new steady state with a change in micro- or meso-climate and the annual receipt of plant residues. The study of changes in the structure of microbial communities seems to be important for understanding the mechanisms underlying the patterns of utilization of different carbon sources.

Experimental Modeling of the Carbon and Nitrogen Cycle Processes

To assess the effect of elevated temperatures on the rate of processes of the carbon and nitrogen cycle were conducted laboratory incubation experiments. Soil samples representing the main horizons of tundra-gley soils, were selected on the various elements of the microrelief, and were incubated for 6 weeks at a temperature from 2°C to 24°C. After the experiment was finished, soil samples were determined: the total content of the SOM and its fractions (Sollins et al. 1999), heterotrophic respiration and ^{13}C in emitted CO_2, as well as $\delta^{13}C$ of microbial biomass (Vance et al. 1987; Paul et al. 1999) and phospholipids fatty acids (PLFA).

The effect of temperature on $\delta^{13}C$ of carbon dioxide emitted. The $\delta^{13}C$ value of the exhaled CO_2 was negatively correlated with temperature in organic horizons ($p < 0.05$), for both bumps and depressions (Figure 1). They linearly increased with -23.9‰ to -26.6‰ for bump soils and of 20.6‰ to -29.8‰ in the depressions in the range from 2°C to 24°C. In the mineral horizons of the $\delta^{13}C$ of exhaled CO_2 is also increased with temperature from 12°C to 24°C. When heterotrophic respiration of soils that have incubated at a temperature of 24°C, was measured at 12°C ("experiment thermal reversion") $\delta^{13}C$ CO_2 shifted to the value close to the one that was found for incubation at 12°C.

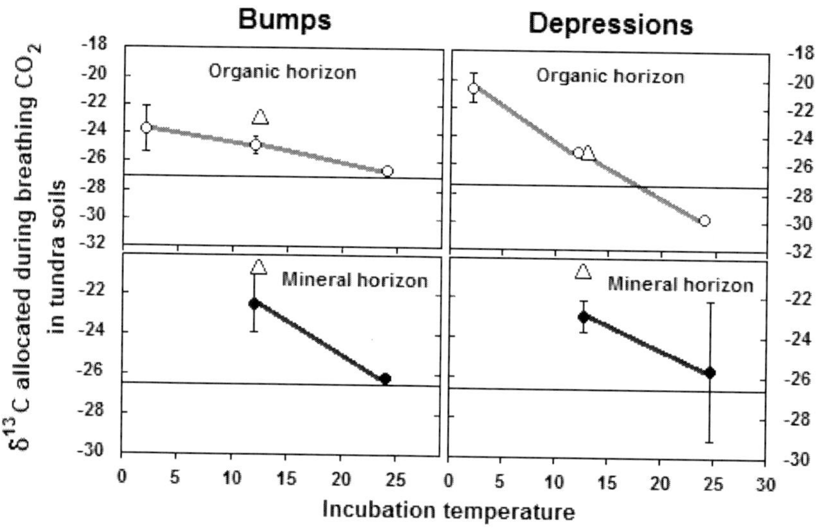

Figure 1. $\delta^{13}C$ allocated during breathing CO_2 in tundra soils incubated for six weeks at 2°C, 12°C and 24°C: transparent and shaded circles are $\delta^{13}C$ of CO_2 respectively for organic and mineral horizons at the temperature of incubation; transparent triangles are $\delta^{13}C$ allocated by breathing CO_2 at 12°C after incubation at 24°C. Values are given as the average ones ± standard deviation (n=3). The horizontal lines show the isotopic characteristics of the original soil.

Radiocarbon dating of CO_2 from soil samples of the depressions showed that the concentration of ^{14}C in CO_2 was at any temperature higher than 100 pMc (percentage of Modern carbon). The concentration of ^{14}C in the organic horizons accounted 119.2, 117.3 and 125.2 pMc, with 2, 12 and 24°C respectively, while in mineral horizons at 12 and 24°C the concentration of ^{14}C amounted to 115.6 and 116.0 pMc. The initial $\delta^{13}C$ value of the soil was close to the $\delta^{13}C$ value of the exhaled CO_2 during incubation temperature 24°C.

The Effect of Temperature on Heterotrophic Respiration

In organic horizons the respiratory rate increased in parallel the whole temperature range (mean Q_{10} of 2.6). Sensitivity to temperature, however, was greater between 2°C and 12°C (Q_{10} – 3.8) than between 12°C and 24°C (Q_{10} – 1.3) (Figure 2).

CO_2 dedicated by heterotrophs in the mineral horizons of the depressions at 2°C was below the detection limit. Between 12°C and 24°C, however, respiratory rate increased (Q_{10} of 2.6), whereas it weakly reduced in bumps (Q_{10} – 0.8). In general, the rate of CO production was similar between the two microsections, and to a greater degree in organic than in mineral horizons.

The Effect of Temperature on Microbial Biomass and Community Structure

Total PLFA used as indicator of microbial biomass, ranged between 1.3 and 4.3 μm PLFA / g SOM, and reaches its highest value at 12°C (Table 1).

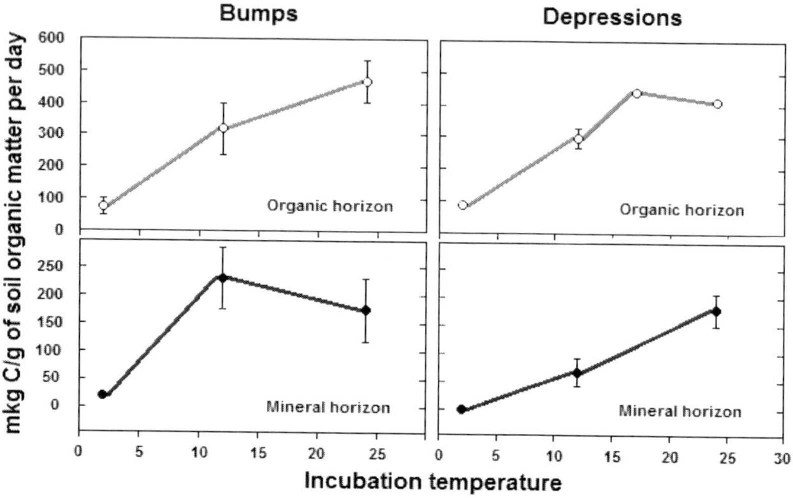

Figure 2. Respiration rate of tundra soil incubated for 6 weeks at 2°C, 12°C and 24°C: values are given as the average ones ± standard deviation (n=3).

Table 1. Microbial biomass (PLFA μm / g SOM) and $\delta^{13}C$ microbial biomass (MB) in tundra soils incubated at different temperatures

	Bumps			Micro-depressions		
	2°C	12°C	24°C	2°C	12°C	24°C
Organic horizon						
MB	1.40	2.93	2.38	1.93	4.27	2.13
	0.11	0.25	0.63	0.16	0.67	0.14
$\delta^{13}C$	-25.94	-26.47	-25.96	-25.00	-24.97	-25.18
	0.1	0.58	0.39	0.47	0.54	0.40
Mineral horizon						
MB	1.90	3.49	2.31	1.26	2.09	1.70
	0.21	0.57	0.28	0.12	0.27	0.12
$\delta^{13}C$	-24.00	-22.81	-23.71	-26.57	-25.55	-25.80
	0.48	0.72	0.29	0.11	0.11	0.22

The $\delta^{13}C$ value of microbial biomass had no significant differences depending on temperature regimes. However, $\delta^{13}C$ values were generally lower in the mineral horizons compared to organic horizons. In both horizons of gram-negative bacteria and fungi were significantly decreased with increasing temperature, while the gram-positive bacteria and saturated with straight-chain PLFA (characterizing both bacteria and fungi) significantly increased (Figure 3) showing significant changes in community structure.

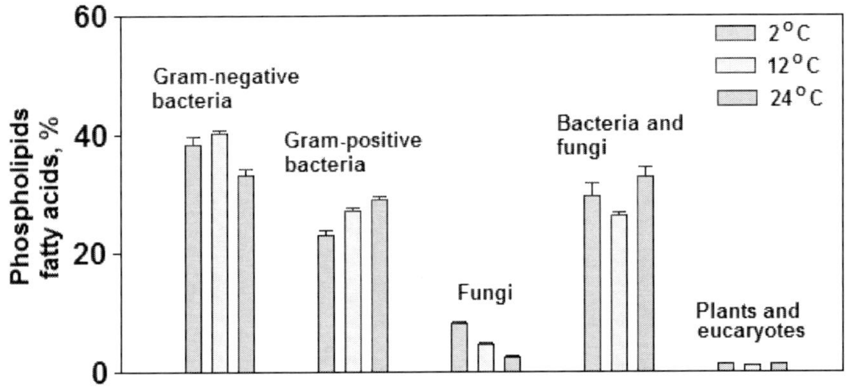

Figure 3. The content of phospholipids fatty acids in the mineral horizons in the incubation experiment.

It was not revealed noticeable effect of temperature on polymerized saturated PLFA with a little chain characterizing plants and other eukaryotes. As the bumps and micro-depressions not shown any noticeable differences in the characteristics of composition of the PLFA data are given in average of both sites.

Soil Fractionation

Chemical fractionation of the soil showed that the different carbon pools exhibit different isotopic characteristics. This principle was generally true for bumps and depressions, as well as for organic and mineral horizons (Table 2).

Table 2. Number and isotopic characterization of organic carbon pools in tundra soils incubated

Bumps						
Carbon fraction	O horizon			A horizon		
	% from C_{tot}	$\delta^{13}C$		% from C_{tot}	$\delta^{13}C$	
Microbial biomass	1.49	-25.89		1.92	-24.89	
	0.14	0.26		0.35	0.41	
Water soluble	1.75	-24.29		3.25	-23.25	
	0.15	0.10		1.23	0.27	
Acid soluble	42.66	-25.19		76.00	-25.54	
	2.72	0.45		40.06	0.21	
Acid unsoluble	51.62	-27.62		15.52	-27.23	
	7.24	0.20		2.46	0.26	
Plant residues	2.47	-27.44		3.31	-27.09	
	0.57	0.27		0.81	0.47	
Initial soil	100.00	-26.50		100.00	-25.77	
Micro-depressions						
Carbon fraction	O horizon		^{14}C (% from modern C)	A horizon		^{14}C (% from modern C)
	% from C_{tot}	$\delta^{13}C$		% from C_{tot}	$\delta^{13}C$	
Microbial biomass	1.49	-26.62	-	2.00	-25.30	-
	0.30	0.32		0.61	0.91	
Water soluble	1.91	-24.94	114.9	3.06	-24.06	114.8
	0.46	0.12		0.25	0.16	
Acid soluble	31.43	-25.13	115.4	37.92	-25.11	117.9
	3.43	0.45		10.58	1.41	
Acid unsoluble	56.76	-28.09	116.4	58.40	-27.05	91.8
	2.80	0.25		6.03	0.87	
Plant residues	8.42	-28.03	-	2.06	-27.92	-
	3.01	0.11		0.27	0.05	
Initial soil	100.00	-27.07	116.0	100.00	-26.20	101.20

The highest value of $\delta^{13}C$ obtained for the water-soluble fraction (-23.2 to -24.9‰), followed by microbial biomass (up -24.9 -26.6‰) and acid fraction (from -25.1 to -25.5‰). Acid soluble fraction had a minimum value of $\delta^{13}C$ (-27.1 to -28.0‰). The original soil was thus enriched in ^{13}C relative to acid soluble fraction and depleted relative to all other factions. As well as for exhaled CO_2 carbon and microbial biomass, the $\delta^{13}C$ values of source soils and fractions of the SOM was lower in the mineral compared to organic horizons. This is consistent with the commonly observed trend of ^{13}C enrichment with increasing soil depth, and indirectly proves the inclusion of microbial components in other (more stable) SOM fractions. Radiocarbon dating of the original soil and the SOM fractions of the depressions showed that the ^{14}C content is also differentiated with depth: mineral horizons are older than organic (Table **2**). The concentration of ^{14}C is about 100 pMc in mineral soils and indicates that these soils are approximately 50 years of age. All soil fractions were younger. Only acid soluble fraction of carbon was significantly older, with a value of 91.8 pMc and radiocarbon age 635 years. In all soils, with the exception of mineral horizons of bumps, acid soluble fraction accounted for the largest share of total organic carbon after acid fraction. In the mineral bump horizons acid fraction is predominant and the minimum carbon content for all the samples was detected in the microbial biomass and water-soluble fraction.

Incubation temperature strongly influenced the value of $\delta^{13}C$ of CO_2 emitted tundra soils. At elevated temperatures the proportion of ^{13}C in CO_2 decreases, whereas at low temperatures CO_2 was more enriched in ^{13}C. Thus, temperature control or affect the use of soil microbe sources of carbon with different $\delta^{13}C$ characteristics or microbial metabolism leads to isotope separation.

Despite the fact that it is hard to imagine that the temperature itself could be the cause of the isotope effect on the individual phases of the enzyme (biochemical) reactions (O'Leary et al. 1992), significant isotope effect could be possible if this changes the metabolism of microorganisms. However, the possible shift of the $\delta^{13}C$ value of the exhaled CO_2 should be balanced by the opposite shift of the $\delta^{13}C$ value of microbial biomass.

In our studies, isotope characteristics of microbial biomass did not correlate with the isotopic characteristics of the exhaled CO_2 therefore the logical conclusion is made about the low probability of such changes of microbial metabolism. Therefore, our data on the decrease in the $\delta^{13}C$ of exhaled CO_2 with increasing temperature most likely indicate the use of different carbon compounds by soil heterotrophs. I.e., soil microbes and tundra ecosystems at elevated temperatures preferably absorb compounds depleted in ^{13}C. Similar results were obtained for forest ecosystems, where they found lower values of $\delta^{13}C$ of CO_2 allocated at 40°C and 22°C than at 4°C (Andrews et al. 2000). It was suggested that at elevated temperatures the more stable fractions of carbon was decomposed by microbes, as previously performed studies have shown that stable carbon compounds were depleted in ^{13}C (Benner et al. 1987). In our studies of tundra soils were also found similar isotopic characteristics of the fractions of carbon. Not extractable, and therefore the most chemically resistant SOM fraction, which mainly contain lignin and humic substances, were the most depleted in ^{13}C, while acid – and water – extractable fractions were relatively more enriched in ^{13}C. Since the value of $\delta^{13}C$ of exhaled CO_2 at 24°C was the most similar to the $\delta^{13}C$ values of the resistant fractions, we concluded that the stable carbon fraction is preferably disposed at 24°C. The $\delta^{13}C$ value of the exhaled CO_2 at 12°C and 2°C are relatively more enriched in ^{13}C, indicating the use of chemically less stable components at low temperatures. In the soils of the depressions, however, values of $\delta^{13}C$ of exhaled CO_2 at 2°C was significantly higher than water-soluble fraction. This can be explained by the fact that each faction SOM itself consists of a variety of (chemically similar) substances, such as soluble humic compounds and sugars (Gregorich et al. 2000), which can have very different isotopic characteristics. Controversial, however, is a judgment on the role of increasing temperature at increasing the use of stable fractions of the SOM or on the rapid depletion of more readily available carbon substrates (Leifeld & Fuhrer 2005). In our research, lowering the temperature to 12°C after incubation at 24°C as a result has increased the $\delta^{13}C$ value of the exhaled CO_2, indicating that the more readily available substrates has not been fully used at 24°C. Therefore, the temperature, undoubtedly, to a greater extent than the availability of substrate influences on selective use of stable carbon of the substrate.

Temperature can increase the solubility of stable substrates, or may destabilize these compounds. Moreover, temperature can affect experimental activity in soils, thus potentially increasing the availability of stable compounds at elevated temperatures. Weintraub (2003) showed that in tundra soils revealed no chemical limitations to decomposition. Instead, it is suggested that the result of the limitations may be adverse environmental conditions such as cold temperatures or high soil moisture, leading to the accumulation of soil organic matter in high latitudes. Radiocarbon dating of soil fractions found that the chemically more stable pools were the oldest, at least in the mineral horizons.

In contrast to other studies (Bol et al. 2003; Waldrop et al. 2004), in our studies, the content of ^{14}C in the exhaled CO_2 has not given specific evidence of the use of older sources of carbon at higher temperatures.

Assuming a constant value of Q_{10} across the temperature gradient, the respiratory rate should exponentially increase with warming (Dalias et al. 2001; Kirschbaum 1995; Lloyd & Taylor 1994). However, in our studies, the value of Q_{10} was not constant and decreased with increasing temperature, although this reduction was insignificant, except for mineral soil mounds. This was unexpected for two reasons: first, in the Arctic ecosystems dominated by organisms adapted to low temperatures (Dalias et al. 2001) and, second, our analyses of exhaled CO_2 suggests that at high temperatures using more resistant carbon compounds. Stable compounds are often considered as passive, i.e., not contributing to heterotrophic respiration (Parton et al. 1987). However, based on our data we believe that the stable carbon pools may be involved in the metabolism of soil microbiota. Thus, the substrate may have only a small effect on the rate of production of CO_2, and the values to which the respiratory rate may decrease as a result of changes in the quality of utilized substrate as a result of warming may be overstated. Various microbial communities can mediate the decomposition of unstable carbon compounds compared to resilient (Schimel et al. 1998; Waldrop et al. 2004). Indeed, we detected a significant change in the composition of the microbial community within the soil as the temperature increased. While the number of fungi and gram-negative bacteria decreased, gram-positive bacteria were relatively more abundant at high temperatures. Also, there was a significant increase among microorganisms with an unbranched chain saturated PLFA. These data are generally consistent with other studies of the development of soil microbial communities to climate warming (Bardgett et al. 1999; Zogg et al. 1997; Waldrop et al. 2004). Apparently, slow-growing fungi are less competitive at high temperatures, while fast-growing gram-positive bacteria are better adapted to these conditions. Thus, in the tundra soils gram-positive bacteria may be responsible for decomposition of more resistant substrates. These organisms, perhaps, are increasing their ability to degrade stable connection by increasing its contribution to experiment (Schimel et al. 2003). In fact, microbial biomass reached a maximum at 12°C, the highest experimental activity (e.g., urease) were detected in the soil incubated at 24°C.

CONCLUSION

Obtained evidence that a stable SOM is preferably decomposed at high temperatures can be extremely important for predicting the impact of climate change on the carbon cycle in the Northern ecosystems. The most important are the conclusions:

1. even the most stable fractions of the SOM can be potentially mineralized by soil microbes and tundra ecosystems;
2. resistant fractions of the SOM can be mineralized in a relatively high degree.

The use of resistant carbon compounds can be explained by changes in the structure of microbial communities dominated by microorganisms with high temperature optimum, which is capable of utilizing more sustainable carbon substrate. It has been proved that temperature, not the depletion of the pool of readily available carbon compounds lead to a preferred degradation of stable fractions of the SOM. Thus, the pool of stable carbon in tundra biome may not be permanent. These SOM values may grow during the cyclic periods of regional temperature increase and may decline in the relevant cold periods.

CONFLICT OF INTEREST

The authors confirm that this article content has no conflicts of interest.

ACKNOWLEDGMENTS

The authors express their deep appreciation and gratitude to their affiliation institutions for the support during research and development.

REFERENCES

Andrews, JA, Matamala, R, Westover, KM, & Schlesinger WH (2000). Temperature effects on the diversity of soil heterotrophs and the delta ^{13}C of soil-respired CO_2. *Soil Biology & Biochemistry*, 32, 699-706.

Benner R, Fogel, ML, Sprague, EK, & Hodson, RE (1987). Depletion of delta ^{13}C in lignin and implication for stable carbon isotope studies. *Nature*, 329, 708-710.

Bardgett, RD, Kandeler, E, Tscherko, D, Hobbs, PJ, Bezemer, TM, Jones, TH, & Thompson, LJ (1999). Below-ground microbial community development in a high temperature world. *Oikos*, 85, 193-203.

Bashkin, V (2016). *Biogeochemical Technologies for Managing Pollution in Polar Ecosystems*. Springer Publishers, Switzerland.

Bashkin, VN, Arno, OB, Arabsky, AK, Barsukov, PA, Priputina, IV, & Galiulin, RV (2012). *Retrospective and Forecast of Geoecological Situation on the Gas Condensate Fields of the Far North*, Gazprom VNIIGAZ Publishing House, Moscow.

Bashkin, VN, & Galiulin, RV (2017). Ecological and biogeochemical footprints of possible climate alterations in polar areas. *Open Ecology Journal* (in this thematic issue).

Bol, R, Bolger, T, Cully, R, & Little, D (2003). Recalcitrant soil organic materials mineralize more efficiently at higher temperatures. *J. Plant Nutr. Soil Sci.,* 166, 300-7.

Chapin, FSI, Hobbie, SE, Bret-Harte, MS, & Bonan, GB (1995a). Causes and consequences of plant functional diversity in arctic ecosystems. In: Chapin, FSI, & Körner, C (Eds). *Arctic and Alpine Biodiversity: Ecological Studies*, 113, Springer-Verlag Publishers, Berlin-Heidelberg, 225-237.

Chapin, FSI, Shaver, GR, Giblin, AE, Nadelhoffer, KJ & Laundre, JA (1995b). Responses of arctic tundra to experimental and observed changes in climate. *Ecology*, 76, 694-711.

Dalias, P, Anderson, JM, Bottner, P, & Couteaux, MM (2001) Temperature responses of carbon mineralization in conifer forests from different regional climates incubated under standard laboratory conditions. *Global Change Biology*, 7, 181-192.

Giardina, CP, & Ryan, MG (2000). Evidence that decomposition rates of organic carbon in mineral soil do not vary with temperature. *Nature*, 404, 858-861.

Gregorich, EG, Liang, BC, Drury, CF, Mackenzie, AF, & McGill, WB (2000). Elucidation of the source and turnover of water soluble and microbial biomass carbon in agricultural soil. *Soil Biology and Biochemistry*, 32, 581-587.

Gu, LH, Post, WM, & King, AW (2004). Fast labile carbon turnover obscures sensitivity of heterotrophic respiration from soil to temperature: a model analysis. *Global Biogeochemical Cycles*, 18, 244-267.

Henry, HAL, & Jefferies RL (2003). Interactions in the uptake of amino acids, ammonium and nitrate ions in the Arctic salt-marsh grass, *Puccinellia phryganodes*. *Plant, Cell & Environment*, 26, 419-428.

Kirschbaum, MUF (1995). Temperature dependence of soil organic matter decomposition, and the effect of global warming on soil organic C storage. *Soil Biology & Biochemistry*, 27, 753-760.

Le Dize, S, Kwiatkowski, BL, Rastetter, EB, Hope, A, Hobbie, JE, Stow, D, & Daeschner, S (2003). Modelling biogeochemical responses of tundra ecosystems to temporal sand spatial variations in climate in the Kuparuk River Basin (Alaska). *J. Geophys. Res.*, 108(D2), doi:10.1029/2001JD000960.

Leifeld, J, & Fuhrer, J (2005). The temperature response of CO_2 production from bulk soils and soil fractions is related to soil organic matter quality. *Biogeochemistry*, 75(3), 433–453.

Liski, J, Ilvesniemi, H, Makela, A, & Westman, CJ (1999). CO_2 emissions from soil in response to climatic warming are overestimated - The decomposition of old soil organic matter is tolerant of temperature. *Ambio*, 28, 171-4.

Lloyd, J, & Taylor, JA (1994). On the temperature dependence of soil respiration. *Functional Ecology*, 8, 315-323.

Luo, YQ, Wan SQ, Hui, DF, & Wallace, LL (2001). Acclimatization of soil respiration to warming in a tall grass prairie. *Nature*, 413, 622-625.

McKane, RB, Rastetter, EB, Shaver, GR, Nadelhoffer, KJ, Giblin, AE, Laundre, JA, & Chapin, FS (1997). Climatic effects on tundra carbon storage inferred from experimental data and a model. *Ecology*, 78 (4), 1170-1187.

Oechel, WC, Hastings, SJ, Vourlitis, GL, Jenkins, M, Riechers, G, & Grulke, N (1993). Recent change of arctic tundra ecosystems from a net carbon dioxide sink to a source. *Nature*, 361, 520-523.

Oechel, WC, Vourlitis, GL, Hastings, SJ, Zulueta, RC, Hinzman, L, & Kane, D (2000). Acclimation of ecosystem CO_2 exchange in the Alaskan Arctic in response to decadal climate warming. *Nature*, 406, 978-981.

O'Leary, MH, Madhavan, S, & Paneth, P (1992). Physical and chemical basis of carbon isotope fractionation in plants. *Plant, Cell and Environment*, 15 (9), 1099-1104.

Parton, WJ, Schimel, DS, Cole, CV, & Ojima, DS (1987). Analysis of factors controlling soil organic matter levels in Great Plains Grassland. *Soil Science Society of America Journal*, 51, 1173-1179.

Paul, EA, Harris, D, Klug, MJ, & Ruess, RW (1999). The determination of microbial biomass. In: Robertson, GP, Coleman, DC, Bledsoe, CS, & Sollins, P (Eds). *Standard Soil Methods for Long-term Ecological Research*, Oxford University Press Publishers, Oxford, 291-317.

Rustad, LE, Campbell, JL, Marion, GM, Norby, RJ, Mitchell, MJ, Hartley, AE, Cornelissen, JHC, Gurevitch, J, & GCTE-NEWS (2001). A meta-analysis of the response of soil respiration, net nitrogen mineralization, and aboveground plant growth to experimental ecosystem warming. *Oecologia*, 126, 543-562.

Shaver, GR, Canadell, J, Chapin, FS, Gurevitch, J, Harte, J, Henry, G, Ineson, P, Jonasson, S, Melillo, J, Pitelka, L, & Rustad, L (2000). Global warming and terrestrial ecosystems: a conceptual framework for analysis. *Bioscience*, 50, 871-882.

Semenov, VM, & Cobut, BM (2015). *Soil Organic Matter*. GEOS Publishing House, Moscow.

Sollins, P, Glassman, C, Paul, EA, Swantson, C, Lajtha, K, Heil, JW, & Elliott, ET (1999). Soil carbon and nitrogen. Pools and fractions. In: Robertson, GP, Coleman, DC, Bledsoe, CS, & Sollins, P (Eds). *Standard Soil Methods for Long-term Ecological Research.* Oxford University Press Publishers, Oxford, 89-105.

Sturm, M, Racine, C, & Tape, K (2001). Climate change - increasing shrub abundance in the Arctic. *Nature*, 411, 546-7.

Townsend, AR, Vitousek, PM, Desmarais, DJ, & Tharpe, A (1997). Soil carbon pool structure and temperature sensitivity inferred using CO_2 and $^{13}CO_2$ incubation fluxes from five Hawaiian soils. *Biogeochemistry*, 38, 1-17.

Vance, ED, Brookes, PC, & Jenkinson DS (1987). An extraction method for measuring soil microbial biomass C. *Soil Biol. Biochem.*, 19, 703–7.

Vourlitis, GL, & Oechel, WC (1999). Eddy covariance measurements of CO_2 and energy fluxes of an Alaskan tussock tundra ecosystem. *Ecology*, 80, 686-701.

Washburn, AL (1956). Classification of patterned ground and review of suggested origins. *Geological Society of America Bulletin*, 67, 170–188.

Weintraub, MN, & Schimel, JP (2003). Interactions between carbon and nitrogen mineralization and soil organic matter chemistry in arctic tundra soils. *Ecosystems*, 6(2), 129-143.

Wookey, PA (2002). Tundra. Encyclopedia of global environmental change. In: Mooney, HA, & Canadell (Eds). *The Earth System: Biological and Ecological Dimensions of Global Environmental Change*, J, Chichester, John, Wiley & Sons Publishers, Ltd. NY, 2, 593-602.

Zogg, GP, Zak, DR, Ringelberg, DB, MacDonald, NW, Pregitzer, KS, & White, DC (1997). Compositional and functional shifts in microbial communities due to soil warming. *Soil Sci. Soc. Am. J.*, 61, 475-481.

In: Ecological and Biogeochemical Cycling … ISBN: 978-1-53612-081-3
Editor: Vladimir N. Bashkin

Chapter 5

BIOPRODUCTIVITY OF TUNDRA ECOSYSTEMS AND NITROGEN UPTAKE BY PLANTS

Vladimir N. Bashkin[1,2], Rauf V. Galiulin[3] and Pavel A. Barsukov[4]

[1]Institute of Physicochemical and Biological Problems of Soil Science of Russian Academy of Sciences, Pushchino, Russian Federation
[2]Gazprom VNIIGAZ LLC, Razvilka, Russian Federation
[3]Institute of Basic Biological Problems of Russian Academy of Sciences, Pushchino, Moscow region, Russian Federation
[4]Institute of Soil Science and Agrochemistry of Siberian Branch of Russian Academy of Sciences, Novosibirsk, Russian Federation

ABSTRACT

At the base of outdoor chamber method the effect of experimental elevated temperature on tundra ecosystem productivity was studied. Furthermore, using ^{15}N techniques, it was found that higher vascular plants of tundra ecosystems, in contrast to more southern ecosystems, have developed an adaptive mechanism to the predominant absorption is not nitrate, but ammonia nitrogen. When comparing the quality of nitrogen-containing substrates such as amino group (glycine), nitrate, ammonium and amide (urea), it was found that all plants uptake the minimum quantity of nitrate nitrogen. The speed of assimilation of ammonium nitrogen by vascular plants was higher on average almost 14 times, amide nitrogen 6 times and the nitrogen of the amino group is almost 3 times than nitrate nitrogen. The speed of assimilation of all forms of nitrogen by mosses and lichens were 3-4 times higher than vascular plants. Spore-bearing plants, as well as vascular, intensely absorb more ammonium nitrogen than nitrate. However, in contrast to vascular plants, the rate of assimilation of the nitrogen of the amino group and the amide nitrogen by mosses and lichens is very high.

Keywords: bioproductivity, tundra ecosystems, nitrogen species, plant uptake, ^{15}N date

INTRODUCTION

It was postulated that the climate changes may alter not only the bioproductivity of tundra ecosystems but also the type of nitrogen regime (Bashkin et al. 2012). In recent times, there are publications that extend our understanding of the role of soluble organic nitrogen compounds (amino acids), previously considered unavailable to plants with mycorrhizae or without it (Kielland 1994; Raab et al. 1996; Falkengren-Grerup et al. 2000; Thornton & Robinson 2005). In parallel, it has been discovered that relatively high concentrations of amino acids present in the soils of Arctic, Alpine and boreal ecosystems (Chapin et al. 1995a; Kielland 1994; Raab et al. 1996; Näsholm et al. 1998; Näsholm & Persson 2001), and in soils of the temperate zone (Murphy et al. 2000; Semenov & Cobut 2015). In the 1990s there were published works demonstrating the ability of Arctic plants to absorb not only nitrate and ammonium nitrogen, but also simple organic compounds - amino acids, urea (Chapin et al. 1995b; Kielland 1994). In respect of urea, it was difficult to prove the direct absorption of this compound, since the surface of plants roots showed a significant activity of the enzyme urease that breaks down urea to ammonium. However, a number of works has been proven at the molecular level, the presence of plant roots systems for absorption as amino acids and urea. In recent years there appeared works showing substantial contribution of soluble organic compounds in nitrogen nutrition of plants in ecosystems with deficiency of mineral nitrogen in the soil. In particular, it is proved that in some ecosystems the relative speed of assimilation of organic nitrogen can be higher than inorganic (Raab et al. 1999), showing a successful competition Arctic plants (the grass bascillica) with soil microorganisms for available amino acids (Henry & Jefferies 2003), and also discovered the ability to absorb a wide range of amino acids not only for Arctic plants, but for typical representatives of the plant communities of the boreal forests (Näsholm & Persson 2001; Bashkin et al. 2012). A comparative study of assimilation of different forms of nitrogen by plants of tundra ecosystems of Northern Siberia has not previously been conducted.

Evaluation of Changes in the Biological Productivity

Methods

In situ study of the effect of elevated temperatures in the processes of the nitrogen cycle and the carbon and productivity of tundra plants was performed using the open hexagonal chambers according to the standard ITEX (International Tundra Experiment). Hexagonal chambers were installed in tundra ecosystems on the Taz Peninsula approximately 5 km from Yamburg gas production point. It is known that in these cameras the average temperature of surface air is increased by an average of 2-3°C in comparison with the adjacent territory, allowing in natural field conditions to investigate the effects of elevated temperatures (Bashkin et al. 2012).

Results

In the most details to date, we have studied the productivity of plants in ecosystems of typical tundra (Figure 1, Tables 1-3).

Table 1 presents the results of determination of the stocks of live and dead plant matter in the dynamics throughout the growing season from June to early September. Sampling was conducted from plots of 0.25 x 0.25 m^2 in 5-fold replication in each date selection. Dead mosses and lichens accounted for nearly half of the inventory of herbal substances in each date selection, which is a consequence of slow decomposition of organic matter in tundra ecosystems due to the short period with positive temperatures, not high temperatures during this period and humidity. Of live plants the greatest contribution to supply plant substances provide lichens (24% on average for all dates) and mosses 12% (average). The maximum biomass of lichens was observed by mid-August, and mosses - in the beginning of the growing season. The contributions of the shrubs *Betula nana* and *Salix sp.* respectively equal to 4.9% and 0.7% of the total mass of vegetable matter. Biomass of shrubs is only 0.4-0.9% of the stock of vegetable substances, and the proportion of herbaceous vegetation is insignificant.

The total stock of plant mass was 2.3 kg/m^2, averaged over all dates selection with a noticeable maximum in the middle of the growing season and two minima in spring and autumn. Given that the stock biomass (mass of living plants) amounted to an average of 43.9% by weight of the total vegetable substances, the total stock of live plants averaged 1.0 kg kg/m^2. For the most common species of shrubs and dwarf shrubs the weight ratio of leaves and stems was determined in dynamics during the vegetation season (Table 2).

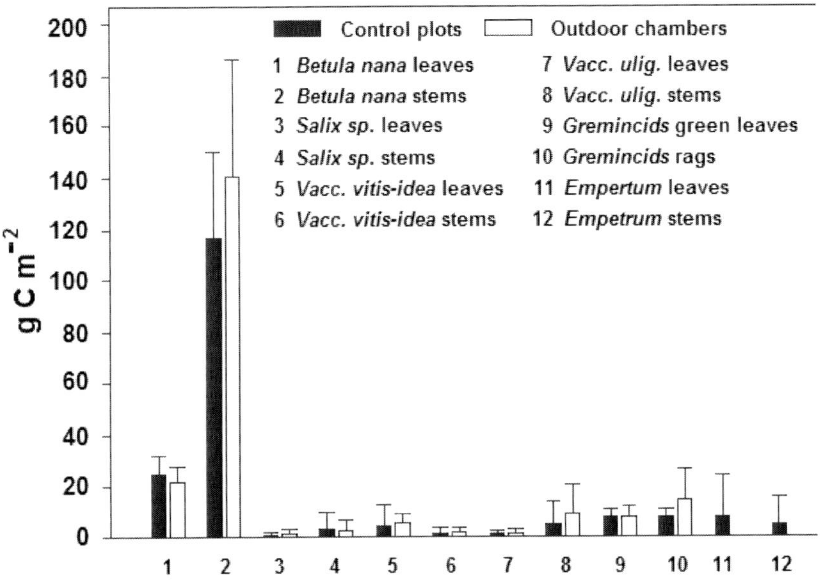

Figure 1. Aboveground biomass of tundra plants in control plots and outdoor chambers.

Table 1. Dynamics of aboveground plant material in typical tundra ecosystem, g/m^2

Biomass type	Sampling date					Mean	% from total mass
	19.06	08.07	29.07	12.08	05.09		
Lichens	524.4	548.2	488.5	675.8	485.3	544.4	23.6
Mosses	370.4	290.1	266.1	240.9	222.9	278.1	12.1
Dead mosses and lichens	1025.8	1314.6	1261.1	1078.5	896.3	1115.3	48.4
Rags	57.8	50.2	39.7	61.7	68.7	55.6	2.4
Litter	70.1	112.2	109.2	160.1	156.0	121.5	5.3
Betula nana - leaves	5.1	29.4	25.6	31.4	8.2	20.0	0.9
Betula nana - stems	60.4	119.9	68.7	111.7	103.3	92.8	4.0
Vaccinium vitis-idea - leaves	8.8	3.8	14.5	6.0	14.4	9.5	0.4
Vaccinium vitis-idea - stems	32.6	2.0	5.6	1.7	5.9	9.6	0.4
Vaccinium uligonosum – leaves	0.4	1.8	2.1	8.2	5.4	3.6	0.2

Table 1. (Continued)

Biomass type	Sampling date					Mean	% from total mass
Vaccinium uligonosum - stems	5.5	5.9	5.0	30.7	32.9	16.0	0.7
Salix sp. - leaves	1.3	1.8	2.7	3.0	8.6	3.5	0.2
Salix sp. - stems	10.5	2.6	8.1	7.7	23.0	10.4	0.5
Ledum decumbens - leaves	3.0	1.1	3.1	1.2	6.3	2.9	0.1
Ledum decumbens - stems	4.7	1.4	7.6	1.2	23.3	7.7	0.3
Calamagrostis lapponum	0.3	1.4	3.9	4.6	5.1	3.0	0.1
Carex ensipholia	4.5	5.7	5.1	8.3	8.7	6.5	0.3
Pedicularis sp.	0.2	1.5	1.8	0.0	0.0	0.7	0.0
Empetrum nigrum	0.0	0.0	5.2	0.0	0.0	1.0	0.0
Eriophorum vaginatum	0.0	0.3	0.5	0.0	0.0	0.2	0.0
Total biomass	2185.8	2494.2	2324.0	2432.8	2074.3	2302.2	100

Table 2. Dynamics of leaf to stem biomass ratio in the typical tundra ecosystem, g/m^2

Species/genera of plants	Sampling date					Mean
	19.06.	08.07.	29.07.	12.08.	05.09.	
Betula nana	0.09	0.25	0.37	0.28	0.08	0.21
Salix sp.	0.12	0.71	0.33	0.40	0.37	0.38
Vaccinium vitis-idea	0.27	1.90	2.59	3.19	1.38	1.87
Vaccinium uligonosum	0.08	0.31	0.43	0.26	0.22	0.26
Ledum decumbens	0.63	0.80	0.41	0.94	0.27	0.61

In the aboveground biomass of shrubs mass of stems is dominated: the ratio leaves/stems is equal on average to 0.25. Different types of shrubs varied significantly according to this index from 0.26 to 1.87. During the growing season, the high ratio of leaves to the mass of stems for different plant species was observed from early July (*Salix sp.*) until the beginning of August (*Vaccinium vitis-idea, Ledum decumbens*).

The concentrations of nitrogen in the vegetal matter of tundra plants vary from 0.3 to 0.9% for mosses and lichens, averaging 0.6 percent (Table 3).

For vascular plants it is even wider variation: from 0.7 to 4.0%. The leaves of all the shrubs contain more nitrogen than stems. The maximum concentrations of nitrogen (an average of 2.3% for the season) were observed for plants of the genus *Pedicularis*, and the minimum (1.2% average) - *Empetrum nigrum*.

Table 3. Dynamics of nitrogen in aboveground plant matter in the typical tundra ecosystem, g/m^2

Biomass type	Sampling date					Mean
	19.06.	08.07.	29.07.	12.08.	05.09.	
Lichens	0.31	0.72	0.42	0.37	0.90	0.54
Mosses	0.58	0.74	0.68	0.62	0.57	0.64
Dead mosses and lichens	0.57	0.73	0.71	0.60	0.70	0.66
Rags	0.82	0.81	0.71	0.68	0.75	0.75
Litter	1.06	1.20	1.14	1.10	0.89	1.08
Betula nana – leaves	4.01	1.88	1.85	1.84	0.79	2.07
Betula nana – stems	0.83	0.77	0.83	0.78	0.82	0.81
Vaccinium vitis-idea - leaves	0.84	1.17	1.01	1.02	0.97	1.00
Vaccinium vitis-idea - stems	0.91	0.98	0.90	0.93	0.85	0.91
Vaccinium uligonosum - leaves	3.37	2.08	3.05	1.71	0.91	2.22
Vaccinium uligonosum - stems	0.84	0.83	0.95	0.93	1.04	0.92
Salix sp. – leaves	3.46	2.45	2.10	1.34	1.22	2.11
Salix sp. – stems	0.71	0.99	0.84	1.50	0.76	0.96
Ledum decumbens - leaves	1.55	1.88	1.79	1.44	3.49	2.03
Ledum decumbens - stems	n.d.	0.84	0.76	0.78	2.73	1.28
Calamagrostis lapponum	2.09	2.68	1.92	1.42	1.05	1.83
Carex ensipholia	2.04	1.79	1.87	1.48	1.13	1.66
Pedicularis sp.	2.57	1.72	1.68	3.04	n.d.	2.25
Empetrum nigrum	n.d.	1.52	0.81	n.d.	n.d.	1.17
Eriophorum vaginatum	n.d.	1.13	1.41	n.d.	n.d.	1.27

Note: n.d. - not determined

Nitrogen Compounds Uptake by Plants of Tundra Ecosystems

Methods

The field experiments were conducted in micro-chambers installed in tundra ecosystems with various local vegetation species. Labeled ^{15}N amino acids, ammonium, nitrate and urea were applied as a source of nitrogen supply. The detailed experiment design was described earlier (Bashkin et al. 2012).

Results

The rate of absorption of the nitrogen of the amino group (glycine), nitrate, ammonium and amide (urea) forms of nitrogen were studied by spore plants and roots of higher plants, and it has varied widely from 0.1 to 29.4 μm N/g of dry matter per hour. All groups of plants in the minimum quantity were absorbed nitrate nitrogen. The speed of assimilation of ammonium nitrogen by vascular plants was higher on average almost 14 times, amide nitrogen 6 times and the nitrogen of the amino group is almost 3 times than nitrate nitrogen (Figure 2, Tables 4-5).

Consequently, higher vascular plants of tundra ecosystems, in contrast to more southern ecosystems, have developed an adaptive mechanism to the predominant absorption is not nitrate, but ammonium since nitrification in tundra soils is inhibited.

The speed of assimilation of all forms of nitrogen by mosses and lichens were significantly higher than roots of vascular plants - in average for the studied substrates 4.1 and 2.6 times, respectively. When examining each substrate separately, the pattern was preserved. Both groups of plants are also more rapidly absorbed ammonium nitrogen than nitrate.

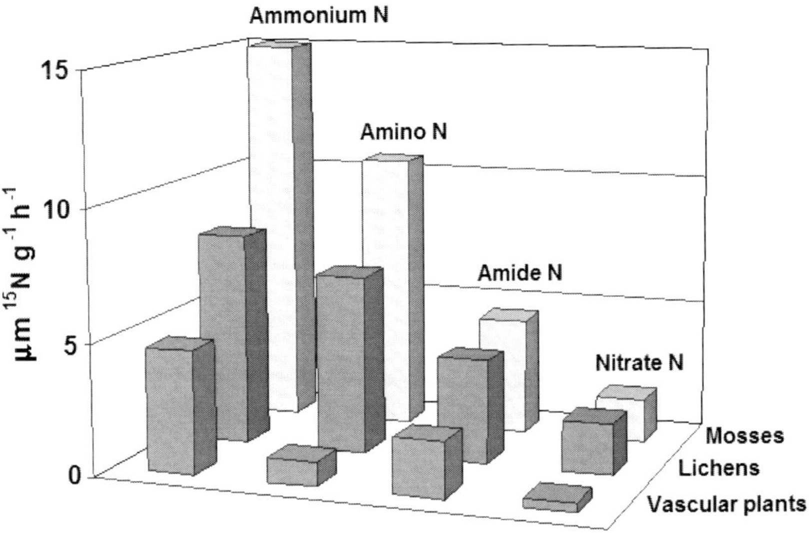

Figure 2. Uptake rate of different forms of nitrogen by plants of southern and typical tundra of the Taz Peninsula.

Table 4. Rate of nitrogen uptake from different substrates by plants of tundra ecosystems in the Taz peninsula, µm ^{15}N h^{-1} g^{-1} (n = 3)

Species/genera of plants	NO_3^-	NH_4^+	Glycine	Urea
Vascular plants				
Salix sp.	0.30	6.12	1.46	3.20
Cassiope sp.	0.10	1.66	0.39	0.63
Carex bigelowii	0.25	3.29	0.85	0.66
Dryas sp.	0.46	6.85	1.51	4.38
Eriophorum sp.	0.48	6.34	1.34	2.79
Valeriana sp.	0.22	4.71	0.27	1.00
Saxifraga sp.	0.58	3.90	0.60	2.01
Mean	0.34	4.70	0.92	2.10
SD	0.06	0.71	0.20	0.54
CV	50	40	57	68
Mosses				
Hylocomium splendens	0.91	10.84	7.18	1.59
Rachmitrium lanuginosum	1.69	11.38	7.61	1.48
Aulacomnium turgidum	1.10	18.57	13.10	3.62
Oncophorus wahlenbergii	1.81	18.52	9.69	2.16

Table 4. (Continued)

Species/genera of plants	NO_3^-	NH_4^+	Glycine	Urea
Polytrichum communis	0.29	4.34	0.52	0.21
Ptilidium ciliare	2.32	17.69	12.34	6.43
Sphagnum aongstroemii	2.98	20.25	18.52	2.03
Sphagnum girgensohnii	2.12	23.75	16.94	2.37
Polytrichum strictum	0.13	6.82	0.21	0.19
Sphagnum girgensohnii	1.54	11.67	8.89	3.84
Ptilidium sp.	2.50	22.24	18.22	22.50
Aulacomnium sp.	2.72	29.39	13.45	7.15
Mean	1.68	16.29	10.55	4.46
SD	0.27	2.13	1.77	1.76
CV	55	45	58	136
Lichens				
Tamnolia verucifera	1.55	7.76	6.53	3.45
Alectoria ochroleuca	2.22	6.79	5.77	1.98
Cladonia sp.	2.52	11.35	8.60	3.76
Cetraria stellate	1.53	5.30	3.04	2.20
Cetraria cuculata	3.07	8.54	9.74	7.26
Cladina rangiferina	2.01	8.37	7.82	3.24
Peltigera aphtosa	0.53	2.45	0.83	0.70
Cetraria islandica	2.31	8.88	9.96	6.98
Nephroma arctica	0.74	10.21	7.33	4.18
Peltigera sp.	2.80	11.35	8.17	5.70
Mean	1.93	8.10	6.78	3.94
SD	0.27	0.87	0.92	0.68
CV	43	34	43	54

However, it is important to note that different groups of plants differed significantly by preference to absorb specific forms of nitrogen. So, if the amount of assimilation of all 4 forms of nitrogen is taken as 100%, then the species of tundra vascular plants preferentially absorb ammonium nitrogen -58%, followed by the amide nitrogen as 26%, nitrogen of amino acids to 11% and the least nitrate - nitrogen 4%. Although mosses and lichens are also preferred ammonia nitrogen - 39-49%, but they are more intensively (then vascular plants) assimilated nitrogen of amino acids - 32-33%, to a lesser extent - amide nitrogen, 14-19% and then nitrate nitrogen, 5-9%.

Thus, the studied spore-bearing plants more preferred organic nitrogen (glycine), compared to vascular plants. In general, the share of influence of the factor "group of plants" responsible for 17% of the variance of the rate of nitrogen assimilation, and the factor "form of nitrogenous substrate" - 32% (calculated from the results of two-factor analysis of variance).

Table 5. Uptake rate of nitrogen from different substrates by plants of southern and typical tundra, the Taimyr and Taz peninsulas, μm $^{15}N\ h^{-1}\ g^{-1}$ (n = 7 ... 12)

	Substrate					LSD_{05}
	NO_3^-	NH_4^+	glycine	urea	mean	
Uptake rate, μm $^{15}N\ h^{-1}\ g^{-1}$						
Vascular plants	0.34	4.70	0.92	2.10	2.01	a
Mosses	1.68	16.29	10.55	4.46	8.25	c
Lichens	1.93	8.10	6.78	3.94	5.19	b
Mean	1.31	9.69	6.08	3.50	-	-
Relative absorption rate (the speed of assimilation of nitrate nitrogen)						
Vascular plants	1	14	3	6	-	
Mosses	1	10	6	3	-	-
Lichens	1	4	4	2	-	-
Mean	1	9	4	4	-	-
Relative absorption rate (% of total)						
Vascular plants	4	58	11	26	-	-
Mosses	5	49	32	14	-	-
Lichens	9	39	33	19	-	-
The coefficients of variation of absolute values of the absorption rate						
Vascular plants	50	40	57	68	54	-
Mosses	55	45	58	136	74	-
Lichens	43	34	43	54	44	-
Mean	49	40	53	86	-	-
The standard error of the absolute values of the absorption rate						
Vascular plants	0.06	0.71	0.20	0.54	-	-
Mosses	0.27	2.13	1.77	1.76	-	-
Lichens	0.27	0.87	0.92	0.68	-	-

Table 6. Rate of absorption of the amide nitrogen by mosses and lichens of the southern tundra in the Taz Peninsula, μm ^{15}N h^{-1} g^{-1} (n = 6)

Plant group	Plant species	Mean	LSD_{05}	CV	SD
Mosses	*Polytrichum strictum*	0.19	a	24	0.03
	Aulacomnium turgidum	3.62	b	13	0.27
Lichens	*Cetraria cuculata*	7.26	c	7	0.30
	Cladina rangiferina	3.24	b	3	0.05

Individual members of tundra phytocenoses showed a large variation in the absorption rate of nitrogen inside the 3 plant groups (vascular plants, mosses and lichens): the coefficients of variation varied from 34 to 136% (see Table 5). The minimum coefficients of variation obtained for lichens (average 44% for all 4 nitrogen substrates), the maximum - moss (74%), vascular plants occupied an intermediate position (54%). For all groups of plants the least variation in the rate of absorption obtained for ammonium nitrogen and highest for the amide. The presence in the sample of members of the same species/genera of plants, but selected from different habitats (with the Taimyr and Taz peninsulas), allowed us to make the assumption that the characteristics of nitrogen nutrition of individual plants is caused not only by genotype, but a number of other factors, among which can be both edaphic conditions and abiotic environmental conditions (temperature, humidity, physicochemical properties of soils, etc.).

Given the high variation of the indices of absorption of the amide nitrogen by mosses, the separate experiment was conducted for comparative study of the assimilation of nitrogen from urea by 2 types of widespread mosses and 2 species of lichens (Table 6), which were selected from the same habitat - the subzone of southern tundra of the Taz Peninsula.

The obtained data confirmed a significantly smaller variation of indicators of absorption rate of the amide nitrogen by lichens (5% average), compared to mosses (19% average).

For the same widespread species of mosses and lichens the kinetic experiments on the effect of rate of nitrogen assimilation of different substrates depending on its concentration in the solution were performed (see Figure 2 and Table 5). For purposes of clarity, data for the highest concentration (5 mm) in the figure are not shown, but illustrated logarithmic regression line takes into account the value of speed and at this concentration. Calculated according to this maximum absorptive capacity of nitrogen shows high values, both for mosses and lichens (Table 7).

It is important to note that for most plants (except *Aulacomnium turgidum*), the maximum absorptive capacity of organic nitrogen (glycine) is even higher than ammonium nitrogen.

As mentioned above, the regional warming will influence the distribution of spore and vascular plants in tundra ecosystems (Chapin et al. 1995a; Sturm et al. 2001). Thus, the abundance and distribution of different groups of plants, in accordance with the identified differences in the rate of absorption of the individual nitrogen-containing substrates will have a significant impact on the nitrogen biogeochemical cycle and the functioning of ecosystems as a whole. Since mosses and lichens efficiently catch nutrients from precipitation and surface waters, which are often enriched by nutrients leached from the soil and plants (Bates 2000), these plants prevent the loss of nutrients from ecosystems.

Table 7. Maximum absorption of nitrogen by mosses and lichens from different substrates, µm ^{15}N h^{-1} g^{-1} (n = 3 for each of the 7 concentrations of ions)

Plant group	Plant species	Substrate	Mean
Mosses	*Polytrichum strictum*	NO_3^-	37
		NH_4^+	31
		glycine	55
	Aulacomnium turgidum	NO_3^-	2
		NH_4^+	68
		Glycine	32
Lichens	*Cetraria cuculata*	NO_3^-	24
		NH_4^+	42
		glycine	44
	Cladina rangiferina	NO_3^-	13
		NH_4^+	20
		Glycine	32

We obtained higher values of the speed of absorption of nitrogen by studied spore plants compared to vascular, and it can be of great importance for the functioning of tundra ecosystems. Especially in spring, during the snowmelt, the dissolved organic and mineral compounds of nitrogen are usually found in abundance in the Arctic ecosystems, mosses and lichens accumulate large reserves of this nutrient. In the spatial dimension this leads to the heterogeneous distribution of nitrogen with the elements of the microrelief, because vegetation of depression contains more spore plants than bumps. Therefore, depression areas are most likely accumulating more nitrogen. Later during the growing season, the nitrogen along with other nutrients, slowly out from spore-bearing plants by leaching (Longton 1988). Thus a constant supply of nutrients is maintained in the depression soil, resulting in a shift of the supply of nutrients from bumps into depression areas. Constantly occurring in the Arctic ecosystems active cryoturbation processes contribute to the mixing of the soil with high and low nitrogen content, as in space (soil surface) and in the soil profile, contributing to a more homogeneous distribution of the nitrogen stock. Given the high content of dissolved organic and mineral nitrogen compounds in surface water after snowmelt and liquid precipitation, mosses and lichens, of course, play a very important role in preventing of losses from tundra ecosystems and in maintenance of a constant stock of this biophilic element at a certain level.

Thus, it was found that higher vascular plants of tundra ecosystems, in contrast to more southern ecosystems, have developed an adaptive mechanism to the predominant absorption is not nitrate, but ammonia nitrogen. When comparing the quality of nitrogen-containing substrates of the nitrogen of the amino group (glycine), nitrate, ammonium and amide (urea) forms of nitrogen, it was found that all plants absorb the minimum quantity of nitrate nitrogen. The speed of assimilation of ammonium nitrogen by vascular plants was higher on average almost 14 times, amide nitrogen 6 times and the nitrogen of the amino group is almost 3 times than nitrate nitrogen. The speed of assimilation of all forms of nitrogen by mosses and lichens were 3-4 times higher than vascular plants. Spore-bearing plants, as well as vascular, intensely absorb more ammonium nitrogen than nitrate. However, in contrast to vascular plants, the speed of assimilation of the nitrogen of the amino group and the amide nitrogen by mosses and lichens is very high. Thus, considering the high content of dissolved organic and mineral compounds of nitrogen in surface waters (especially after snowmelt), mosses and lichens due to the rapid assimilation of nitrogenous substances undoubtedly play a very important role in the prevention of nitrogen losses from tundra ecosystems.

For typical representatives of phytocenosis of the subzone of typical tundra (the Gydan Peninsula) we studied the dynamics of accumulation of vegetable matter and the nitrogen stock in the biomass during the vegetation season. The regularities of the distribution of carbon and nitrogen in typical tundra ecosystems of the main elements of microrelief were revealed. The nitrogen stock in the soil exceeds the nitrogen stock in the biomass of plants in bumps by 18 times, in depressions -28 times, in mounds of swelling - to 9-fold. The stock of nitrogen in plants of depressions contain maximum amount of nitrogen that is greater than the amount of nitrogen in the plants of bumps in 1.2 times, and mounds of swelling -1.8 times. The stock of nitrogen in the roots for all elements of the microrelief exceeds the stock of nitrogen in the aboveground parts of plants: 1.2 times for depressions, 5 times for bumps and 10 times for mounds of swelling.

Analysis of isotopic composition of carbon and nitrogen was carried out in plants and pools of nonliving plant matter of selected ecosystems of the typical tundra. Table 8 presents the results of these analyses, expressed using $\delta^{15}N$ and $\delta^{13}C$, which are commonly used for the characteristics of the fluctuations of the isotopic composition, respectively, of the nitrogen and carbon of various natural compounds/objects.

Differences in natural enrichment usually vary in the range of ten units $\delta^{15}N$ of atmospheric nitrogen and several dozen $\delta^{13}C$ units of atmospheric carbon. For example, one nitrogen $\delta^{15}N$ units = 0.000365 atom. % ^{15}N, as the natural enrichment of the standard (atmospheric nitrogen) is 0.365 atom. % ^{15}N.

Accordingly these data can be considered as the relevant database on the natural variation of the isotopic composition of carbon and nitrogen for most typical representatives of a number of tundra plants and pools of nonliving vegetable matter in the typical tundra ecosystem.

Table 8. Isotopic composition of nitrogen and carbon in various plants and dead vegetable matter in the ecosystem of the typical tundra

Plant species	Element	Parameters	Sampling dates				
			19.06.	08.07.	29.07.	12.08.	05.09.
Betula nana - leaves	$\delta^{15}N$	m	-6.8	-5.3	-6.1	-7.6	-5.6
		SD	0.8	0.6	0.9	2.0	0.9
	$\delta^{13}C$	m	-28.5	-29.4	-30.1	-29.9	-30.3
		SD	0.3	0.6	0.4	0.6	0.7
Betula nana – stems	$\delta^{15}N$	m	-6.7	-6.2	-6.6	-6.9	-5.5
		SD	1.0	1.0	0.6	2.0	2.4
	$\delta^{13}C$	m	-29.1	-29.5	-29.9	-29.9	-29.7
		SD	1.3	0.7	0.5	0.2	1.6
Salix sp. - leaves	$\delta^{15}N$	m	-3.6	-3.7	-3.4	-0.8	0.6
		SD	0.5	-	1.1	5.2	3.7
	$\delta^{13}C$	m	-25.8	-27.2	-27.9	-28.6	-28.0
		SD	0.3	-	0.8	1.4	1.7
Salix sp. - stems	$\delta^{15}N$	m	-4.2	-4.3	-3.6	-4.2	-4.1
		SD	0.1	-	1.3	0.9	0.3
	$\delta^{13}C$	m	-27.0	-27.2	-27.8	-28.9	-28.4
		SD	0.6	-	0.5	0.3	0.4
Vaccinium vitis-idea - leaves	$\delta^{15}N$	m	-5.4	-5.4	-5.1	-6.4	-6.4
		SD	0.4	0.7	1.2	2.9	2.0
	$\delta^{13}C$	m	-29.8	-30.1	-29.9	-30.3	-30.1
		SD	0.9	0.2	0.4	0.3	0.5
Vaccinium vitis-idea – stems	$\delta^{15}N$	m	-5.9	-5.6	-5.8	-8.4	-6.9
		SD	0.7	0.6	0.8	5.6	1.4
	$\delta^{13}C$	m	-17.5	-29.1	-29.0	-29.5	-30.0
		SD	26.2	0.4	0.5	0.6	0.4
Vaccinium uligonosum - leaves	$\delta^{15}N$	m	-4.8	-4.7	-4.4	-5.1	-4.9
		SD	1.0	0.9	1.7	2.3	1.8
	$\delta^{13}C$	m	-28.2	-30.2	-30.2	-30.3	-30.4
		SD	0.3	0.3	1.6	0.5	1.2
Vaccinium uligonosum – stems	$\delta^{15}N$	m	-5.1	-5.5	-5.8	-6.8	-5.1
		SD	0.8	0.6	0.7	3.0	0.8
	$\delta^{13}C$	m	-29.6	-29.3	-30.2	-29.9	-30.6
		SD	0.6	0.8	0.6	0.4	0.7

Plant species	Element	Parameters	Sampling dates				
Calamagrostis lapp.	$\delta^{15}N$	m	2.1	1.1	2.0	2.2	-3.6
		SD	0.6	1.7	0.5	0.4	5.5
Calamagrostis lapp.	$\delta^{13}C$	M	-26.9	-25.0	-26.0	-26.7	-28.7
		SD	1.7	2.8	1.3	1.0	1.9
Carex ensifolia	$\delta^{15}N$	m	2.5	1.8	3.0	1.0	2.0
		SD	0.5	0.5	0.7	2.3	0.9
	$\delta^{13}C$	m	-26.7	-26.7	-26.9	-26.0	-27.4
		SD	0.6	0.5	0.5	2.8	0.9
Pedicularis	$\delta^{15}N$	m	-7.1	-6.2	-7.6	-3.4	n.d.
		SD	-	0.7	-	-	-
	$\delta^{13}C$	m	-27.7	-28.2	-28.9	-28.1	n.d.
		SD	-	0.3	-	-	-
Mosses	$\delta^{15}N$	m	-2.6	-3.8	-1.3	-0.8	-3.3
		SD	0.7	1.4	1.4	5.4	1.3
	$\delta^{13}C$	m	-28.2	-28.2	-27.9	-28.5	-27.9
		SD	0.3	0.6	0.4	0.6	1.5
Lichens	$\delta^{15}N$	m	-3.7	-2.2	-2.6	-2.8	-0.9
		SD	0.6	1.9	0.8	1.7	4.0
	$\delta^{13}C$	m	-26.3	-26.3	-25.8	-25.3	-26.9
		SD	0.4	0.4	0.4	2.2	0.6
Dead lichens and mosses	$\delta^{15}N$	m	-1.2	-2.4	-1.7	-1.7	-1.5
		SD	1.9	0.6	0.9	2.3	2.3
	$\delta^{13}C$	m	-27.0	-26.9	-26.7	-26.8	-28.1
		SD	0.4	0.6	0.3	0.5	0.7
Rags	$\delta^{15}N$	m	-0.3	-0.2	0.3	1.7	0.8
		SD	0.9	1.3	3.3	2.2	3.5
	$\delta^{13}C$	m	-26.7	-26.7	-26.7	-26.3	-27.0
		SD	0.5	0.7	1.9	0.5	0.3
Litter	$\delta^{15}N$	m	-5.0	-3.6	-3.9	-4.1	-1.1
		SD	0.6	2.3	0.4	1.2	2.0
	$\delta^{13}C$	M	-29.6	-29.1	-29.7	-29.7	-28.6
		SD	0.6	1.7	0.2	0.4	1.9

Note: m - average; SD - standard deviation, where SD are not given - the results are presented with one repetition; n.d. - not determined

CONCLUSION

It was found that higher vascular plants of tundra ecosystems, in contrast to more southern ecosystems, have developed an adaptive mechanism to the predominant absorption is not nitrate, but ammonia nitrogen. When comparing the quality of nitrogen-containing substrates of the nitrogen of the amino group (glycine), nitrate, ammonium and amide (urea) forms of nitrogen, it was found that all plants absorb the minimum quantity of nitrate nitrogen. The speed of assimilation of ammonium nitrogen by vascular plants was higher on average almost 14 times, amide nitrogen 6 times and the nitrogen of the amino group is almost 3 times than nitrate nitrogen. The speed of assimilation of all forms of nitrogen by mosses and lichens were 3-4 times higher than vascular plants. Spore-bearing plants, as well as vascular, intensely absorb more ammonium nitrogen than nitrate.

For the first time for most typical representatives of a number of tundra plants and pools of nonliving vegetable matter in the typical tundra ecosystem there were obtained the data on the natural variation of the isotopic composition of carbon and nitrogen.

Acknowledgments

The authors express their deep appreciation and gratitude to their affiliation institutions for the support during research and development.

Conflict of Interest

The authors confirm that this article content has no conflicts of interest.

References

Bates, JW (2000). Mineral nutrition, substratum ecology, and pollution. In: Shaw, AJ, and Goffinet, B (Eds). *Bryophyte Biology*. Cambridge Univ. Press Publishers, Cambridge, 248-311.

Bashkin, VN, Arno, OB, Arabsky, AK, Barsukov, PA, Priputina, IV, and Galiulin, RV (2012). *Retrospective and Forecast of Geoecological Situation on the Gas Condensate Fields of the Far North*. Gazprom VNIIGAZ Publishing House, Moscow.

Chapin, FSI, Hobbie, SE, Bret-Harte, MS, and Bonan, GB (1995a). Causes and consequences of plant functional diversity in arctic ecosystems. In: Chapin, FSI, and Körner, C (Eds.). *Arctic and Alpine Biodiversity: Ecological Studies*, 113, Springer-Verlag Publishers, Berlin-Heidelberg, 225-237.

Chapin, FSI, Shaver, GR, Giblin, AE, Nadelhoffer, KJ, and Laundre, JA (1995b). Responses of arctic tundra to experimental and observed changes in climate. *Ecology*, 76, 694-711.

Falkengren-Grerup, U, Månsson, KF, and Olsson, MO (2000). Uptake capacity of amino acids by ten grasses and forbs in relation to soil acidity and nitrogen availability. *Environmental and Experimental Botany,* 44, 207-219.

Henry, HAL, and Jefferies, RL (2003). Interactions in the uptake of amino acids, ammonium and nitrate ions in the Arctic salt-marsh grass, *Puccinellia phryganodes*. *Plant, Cell and Environment*, 26, 419-428.

Kielland, K (1994). Amino acid absorption by arctic plants: implications for plant nutrition and nitrogen cycling. *Ecology*, 75, 2373-2383.

Longton, RE (1988). *Biology of Polar Bryophytes and Lichens: Studies in Polar Research*. Cambridge Univ. Press Publishers, Great Britain.

Murphy, DV, Macdonald, AJ, Stockdale, EA, Goulding, KWT, Fortune, S, Gaunt, JL, Poulton, PR, Wakefield, JA, Webster, CP, and Wilmer, WS (2000). Soluble organic nitrogen in agricultural soils. *Biology and Fertility of Soils*, 30, 374-387.

Näsholm, T, Ekblad, A, Nordin, A, Giesler, R, Högberg, M, and Högberg, P (1998). Boreal forest plants take up organic nitrogen. *Nature*, 392, 914-916.

Näsholm, T, and Persson, J (2001). Plant acquisition of organic nitrogen in boreal forests. *Physiol. Plant.*, 111, 419-426.

Raab, TK, Lipson, DA, and Monson, RK (1996). Nonmycorrhizal uptake of amino acids by roots of the alpine sedge *Kobresia myosuroides*: implications for the alpine nitrogen cycle. *Oecologia*, 108, 488-494.

Semenov, VM, and Cobut, BM (2015). *Soil Organic Matter*. GEOS Publishing House, Moscow.

Thornton, B, and Robinson, D (2005). Uptake and assimilation of nitrogen from solutions containing multiple N sources. *Plant, Cell and Environment*, 28, 813-821.

PART II. BIOGEOCHEMICAL TECHNOLOGIES BASED ON ECOLOGICAL AND BIOGEOCHEMICAL CYCLING

In: Ecological and Biogeochemical Cycling ... ISBN: 978-1-53612-081-3
Editor: Vladimir N. Bashkin

Chapter 6

ACTIVITY OF CATALASE AND DEHYDROGENASE ENZYMES AS A TOOL FOR CONTROLLING THE SOIL RECLAMATION PROCESS

Rauf V. Galiulin[1] and Vladimir N. Bashkin[2,3]
[1]Institute of Basic Biological Problems of Russian Academy of Sciences, Pushchino, Russian Federation
[2]Institute of Physicochemical and Biological Problems of Soil Science of Russian Academy of Sciences, Pushchino, Russian Federation
[3]Gazprom VNIIGAZ LLC, Razvilka, Russian Federation

ABSTRACT

The activity of catalase and dehydrogenase enzymes was used as a tool for controlling reclamation process of physically disturbed soils of the Taz Peninsula (The Yamalo-Nenets Autonomous Okrug, north of West Siberia, Russia) in the impacted ecosystems of a gas production company. Regeneration of a vegetable cover as self-restoration of the disturbed soils was diagnosed by the catalase activity analysis, and efficiency of peat based reclamation of disturbed soils was diagnosed by the dehydrogenase activity analysis.

Keywords: physically disturbed soil, reclamation, regeneration of vegetative cover, peat use, activity of catalase and dehydrogenase

INTRODUCTION

From the geoecological point of view, process of soil reclamation represents the reconstruction of its fertility lost due to various reasons to which it is possible to refer productive activity of the gas industry in the conditions of the Far North. This activity consists of conducting exploration works, arrangement and development of new fields, gas and gas condensate production and their preparation for transport. Very often this is accompanied with mechanical impacts on a soil-vegetable cover when tundra soils partially or completely lose a vegetable cover and an organogenic layer, and the mineral horizons come to a day surface. Nevertheless the soil has self-restoration in the form of regeneration of a vegetable cover, originally settling by the different types of grass-cereal association not characteristic of the natural communities which subsequently are forced out by native for a tundra zone - mosses and lichens. However in the conditions of severe climate of the tundra about self-restoration of physically disturbed soils it will be possible surely to judge only decades later after the beginning of this process (Vasilevskaya & Kirilishin 1993, Vasilevskaya & Grigoriev 2002). Therefore, a local peat can be applied for stimulating the reclamation of physically disturbed soils in the northern impacted ecosystems. Furthemore, we need the express assessment of this process efficiency in the field conditions.

Peat applying for physically disturbed soil reclamation is due to some reasons. First, this organogenic material consists both of partly decayed plant remains and specific products of their degradation, humus and humic acids, which are characterized by the high content of nutrients, especially nitrogen, up to 3.5%. Secondary, peat possesses a certain pool of various physiological groups of the microorganisms (ammonifiers-aerobes, sporous bacteria, oligonitrofila, fungi, nitrifiers, denitrifiers, butyric-acid bacteria) participating in degradation of peat organic substance that makes available nutrient elements for plants (Emelyanova & Kramarenko, 2004)

As for key indicators of reclamation process, it is possible to refer activity of soil enzymes. This indicator differs in high sensitivity and expressivity, sufficient accuracy and stability of indicators in comparison with the existing methods of microbiological analysis, such as direct calculation of microorganisms of different physiological groups, determination of quantity of microorganisms on different nutrient media and also their biomass (Babieva & Zenova 1989).

The specified geoenvironmental problem was solved on the example of the disturbed tundra soils of the Taz Peninsula (The Yamalo-Nenets Autonomous Okrug, north of West Siberia, Russia) as region of productive activity of 100% subsidiary company of Gazprom - Gazprom Dobycha Yamburg LLC. As known the Taz Peninsula is in the north of the West Siberian Plain between the Ob Bay in the west and the Taz Bay in the east (Figure 1). The territory of the peninsula on a relief flat, is covered with lakes and swamps and has vegetation characteristic of the tundra, and is used by local population for pasturable reindeer breeding. Researches of the last years showed existence of a problem of physically disturbed tundra soils on the Taz Peninsula as a result of productive activity of Gazprom Dobycha Yamburg LLC that directly or indirectly can affects in conducting pasturable reindeer breeding (Galiulin et al., 2013, 2014, 2016, Bashkin et al., 2015).

In this regard the main objective of work consisted in stimulation of the reclamation of disturbed tundra soils by means of local peat and an express assessment of this process efficiency in the conditions of *in vitro* with use a catalase and dehydrogenase enzyme activity as tool of controlling of the soil reclamation process on the Taz Peninsula.

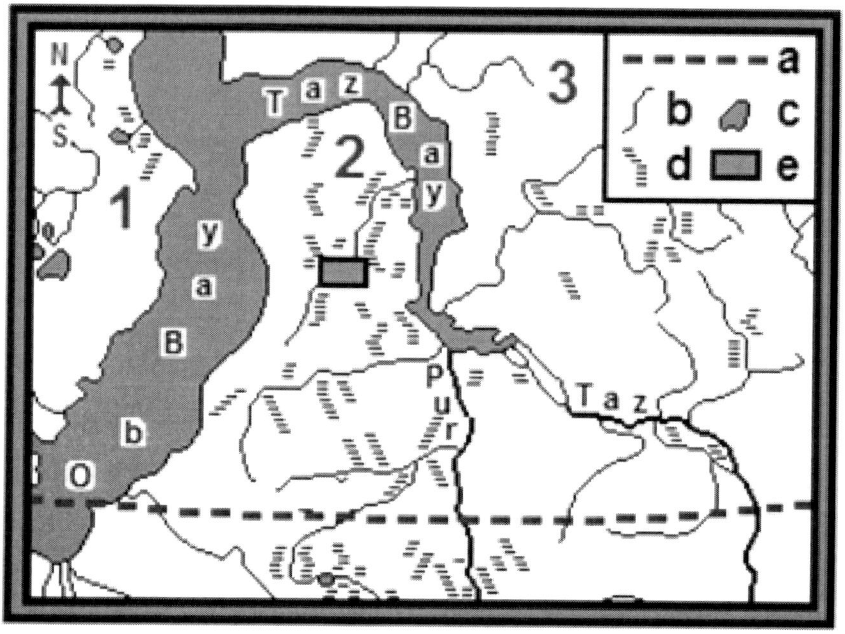

Figure 1. Map-scheme of the territory under study. 1 – The Yamal Peninsula; 2 – The Taz Peninsula (68°09' N, 76°02' E); 3 – The Gydan Peninsula; a – The North Polar Circle, b – rivers, c – lakes, d - swamps, e – region of disturbed soils reclamation.

Materials and Methods

Average samples from 0-6 cm layer of two physically disturbed tundra soils representing the consolidated sand on granulometric composition were selected on sites around of location an installations of the complex gas preparation providing collection and handling of natural gas and gas condensate according to requirements of industry and state standards. On one site the vegetable cover was absent, on other site the regeneration of a vegetable cover in the form of certain species of grass-cereal association, as well as mosses was noted. The content of organic carbon in soil without vegetable cover was 0.2%, and with a vegetable cover – 0.9%. In experiences for soil reclamation, used the local peat with 54% ash-content added to the soil in the ratio 1:4. As a standard 0-10 cm layer of peaty-gleezem typical tundra with 39% ash-content was used (Khrenov 2011). The short physicochemical characteristic of the studied samples of soils and peat is given in Table 1.

Table 1. Physicochemical characteristic of soil and peat samples from the Taz Peninsula

Sample	Density, g/cm^3	Capillary moisture capacity, %	Full moisture capacity, %	pH_{H2O}
Peaty-gleezem typical tundra	0.39	216	315	5.1
Peat	0.40	216	420	5.4
Disturbed soil without vegetative cover	1.71	25	32	5.9
Disturbed soil without vegetative cover + peat, 4:1	1.04	66	80	5.1
Disturbed soil with vegetative cover	1.54	37	43	5.1
Disturbed soil with vegetative cover + peat, 4:1	0.89	78	102	5.3

For assessment of the soil reclamation efficiency, 50 g samples without addition and with peat addition, humidified to 70% of a full moisture capacity incubated in Petri's cups in the thermostat at a temperature 30 °C. In dynamics i.e., on 5, 10, 20 and 30 days analyzed a catalase and dehydrogenase activity of samples from various experimental variants by the methods protected by patents of the Russian Federation (Bashkin et al., 2010, 2010a).

Results and Discussion

It is known that from the studied enzymes the catalase initiates degradation reaction of the hydrogen peroxide to water and oxygen, which is formed in the course of breath of soil organisms and as a result of various biochemical reactions of organic substances oxidation. The catalase activity of soils can be considered not only as an indicator of functional activity of microflora, but also safety of enzymes in postmortal plant material (Efremova & Ovchinnikova 2008). As for the dehydrogenase, its role consists in dehydrogenation reaction catalysis, i.e., hydrogen splitting off, of the organic substances (carbohydrates, alcohols and organic acids) arriving with the vegetable remains. The regression analysis has revealed a high positive associativity (determination coefficient, r^2 = 0.90-098) of catalase and dehydrogenase activity as linear function that consists in transfer of the hydrogen splitting off by dehydrogenase to air oxygen with formation of water and (or) hydrogen peroxide which subsequent degradation is carried out by a catalase.

The data of Table 2 show that closest to catalase activity of peaty-gleezem typical tundra (taken for 100%), was the corresponding activity of disturbed soil with a vegetable cover, which during observation made 61.5-75.0% that is higher than activity even of the peat used for soil reclamation.

It allows to draw a conclusion on a significant contribution of plants to soil biochemical activity through postmortal plant material, owing to the enzymes remaining in its. Apparently regeneration of a vegetable cover as self-restoration of the disturbed soil was diagnosed by increase in catalase activity. The opposite picture was observed in case of dehydrogenase activity correlating with total quantity of microorganisms in the soil. Here we can refer to previous studies, when increasing both enzyme activity and functional activity of microflora was shown (Efremova & Ovchinnikova 2008). At the same time the microflora enzymes influence on dehydrogenation substrata in the form of carbohydrates, alcohols and organic acids.

Table 2. Dynamics of enzyme activity of the physically disturbed soils from the Taz Peninsula at their reclamation by means of peat

Sample	Catalase activity / Dehydrogenase activity, % from activity of peaty-gleezem typical tundra			
	Day			
	5	10	20	30
Peat	43.8 / 128.5	40/100	53.8 / 96.7	63.6 / 101.2
Disturbed soil without vegetative cover	25.0 / 9.2	20.0 / 6.7	23.1 / 7.6	27.3 / 5.9
Disturbed soil without vegetative cover + peat, 4:1	25.0 / 32.6	20.0 / 23.9	23.1 / 19.0	27.3 / 23.7
Disturbed soil with vegetative cover	75.0 / 18.6	73.3 / 13.6	61.5 / 15.1	72.7 / 11.9
Disturbed soil with vegetative cover + peat, 4:1	37.5 / 70.5	40.0 / 58.6	61.5 / 59.9	54.5 / 41.0

So, for the entire period of observation, peat addition significantly increased the dehydrogenase activity of disturbed soil without vegetable cover and, especially with a vegetable cover. Apparently efficiency of the disturbed soils reclamation by means of peat was diagnosed by increase in dehydrogenase activity. As for dehydrogenase activity of peat, it not only reached the corresponding activity of peaty-gleezem typical tundra, but also in the first 5 days was 28.5% higher.

As the proof of adequacy of dehydrogenase activity use in the time of assessment of the disturbed soils reclamation efficiency by means of peat, the results of correlation and regression analysis of experimental data are served. So, calculation of the coefficient of correlation indicating the direction and degree of an associativity in signs variability has shown existence of strong essential correlation dependence between a dehydrogenase activity and density of various samples (correlation coefficient, r = - 0.95), a dehydrogenase activity and capillary moisture capacity (r = 0.95), and also a dehydrogenase activity and full moisture capacity (r = 0.95) is shown. The corresponding formulas of correlation dependence, i.e., the equations of linear regression allow judging how the productive sign (y) quantitatively changes at change of a factorial sign (x) on a unit of measure had the following appearance:

$$y = 76.9 - 44.4x; y = 2.74 + 0.28x; y = 7.71 + 0.15x$$

As it has appeared, the less density of samples and respectively more their capillary and full moisture capacity determined substantially by an organic component of the used samples, the dehydrogenase activity is higher. It is confirmed by data of previous works (Efremova & Ovchinnikova 2008, Pavliuchuk et al., 1982), where increase in dehydrogenase activity with increase of humidity in various soils was noted. The leading value of humidity in soil dehydrogenase activity is connected with the fact that moisture defines a normal physiological state of the microorganisms and plants producing enzymes in the soil and also supports enzymes and their substrata (carbohydrates, alcohols and organic acids) in a reactionary state.

CONCLUSION

Thus, the activity of catalase and dehydrogenase enzymes can be considered a tool for controlling soil reclamation process. The relevant results of laboratory experiments that were aimed to peat based remediation of the physically disturbed tundra soils of the Taz Peninsula showed that catalase and a dehydrogenase activity analysis are fitting for express assessment of reclamation processes.

CONFLICT OF INTEREST

The authors confirm that this article content has no conflict of interest.

ACKNOWLEDGMENTS

The authors express their deep appreciation and gratitude for their affiliation institutions, namely Institute of Basic Biological Problems of Russian Academy of Sciences and also Institute of Physicochemical and Biological Problems of Soil Science of Russian Academy of Sciences for the support during research and development.

REFERENCES

Babieva, I. P. & Zenova, G. M. (1989). *Biology of Soils*. Moscow State University Publishing House, Moscow, p. 336.

Bashkin, V. N., Bukhgalter, E. B., Galiulin, R. V., Konyaev, S. V., Kalinina, I. E. & Galiulina, R. A. (2010). *Control Method for Cleaning of Soils Contaminated by Hydrocarbons and Neutralization of Hydrocarbons Sludge by Catalase Activity Analysis*. RU Patent 2387995.

Bashkin, V. N., Bukhgalter, E. B., Galiulin, R. V., Konyaev, S. V., Kalinina, I. E. & Galiulina, R. A. (2010a). *Control Method for Cleaning of Soils Contaminated by Hydrocarbons and Neutralization of Hydrocarbons Sludge by Dehydrogenase Activity Analysis*. RU Patent 2387996.

Bashkin, V. N., Galiulin, R. V., Galiulina, R. A., Arno, O. B. & Arabsky, A. K. (2015). Diagnostics of recultivation potential of disturbed soils around extraction of natural gas. *Neftegaz.RU*, 3, 52-4.

Efremova, T. T. & Ovchinnikova, T. M. (2008). Seasonal oxidoreductase activity of drained peat soils in connection with hydrothermal conditions of the environment. *Siberian Ecological Journal*, 3 (XV), 441-9.

Emelyanova, T. Ya. & Kramarenko, V. V. (2004). Substantiation of studying technique of peat deformation properties taking into account change of its degradation extent. *News of the Tomsk Polytechnical University*. 5 (307), p. 54-7.

Galiulin, R. V., Bashkin, V. N., Galiulina, R. A., Priputina, I. V. & Arabsky, A. K. (2013). Recultivation of disturbed tundra soils of the Taz Peninsula by means of peat: an efficiency assessment by means of enzyme activity analysis. *Agrochemistry*, 4, 76-80.

Galiulin, R. V., Bashkin, V. N., Arabsky, A. K. & Galiulina, R. A. (2014). Solution of problem of mechanically disturbed soils at extraction of natural gas. *Journal of Oilgas Construction*, 1, 56-8.

Galiulin, R. V., Galiulina, R. A., Bashkin, V. N. & Arabsky, A. K. (2016). Disturbed tundra soils: granulometry and recultivation. *Neftegaz.RU*, 5-6, 102-5.

Khrenov, V. Ya. (2011). *Soils of Cryolithozone West Siberian: Morphology, Physicochemical Properties, Geochemistry*. Nauka Publishing House, Novosibirsk, p. 211.

Pavliuchuk, Z., Aseeva, I. V. & Sudnitsyn, I. I. (1982). Influence of potential (pressure) of soil moisture on dehydrogenase activity in soils. *Bulletin of the Moscow University, Series 17, Soil Science*, 2, 60-3.

Vasilevskaya, V. D. & Kirilishin, V. V. (1993). Anthropogenic disturbances of soil cover in southern tundra of the Yamal and measures for their prevention. *Bulletin of the Moscow University, Series 17, Soil Science*, 4, 3-9.

Vasilevskaya, V. D. & Grigoriev, V. Ya. (2002). Biological indicators of degradation and self-restoration of soil-vegetable cover of tundra. *Siberian Ecological Journal*, 3 (IX), 355-370.

In: Ecological and Biogeochemical Cycling … ISBN: 978-1-53612-081-3
Editor: Vladimir N. Bashkin

Chapter 7

NMR C13 MEASUREMENTS FOR ASSESSING HUMATES FROM TUNDRA PEATS

***Andrey O. Alekseev*[1] *and Vladimir N. Bashkin*[1,2]**
[1]Institute of Physicochemical and Biological Problems of Soil Science RAS, Pushchino, Russia
[2]Gazprom VNIIGAZ LLC, Razvilka, Russia

ABSTRACT

This paper presents the applicability of the method of 13C NMR spectroscopy to assess the preparations of humic substances isolated from peat in the gas industry impacted ecosystems. There were conducted extraction of humates from these peat samples and performed analytical work to identify the structures of isolated humatic compounds using modern analytical methods including NMR spectroscopy and IR spectroscopy. It is shown that quantitative 13C NMR spectroscopy is one of the most informative methods for physico-chemical studies of the chemical nature of humic acids.

Keywords: 13C NMR spectroscopy, humate, peat, tundra ecosystems, gas industry impacts

INTRODUCTION

The emerging modern global problem of the targeted regulation of imbalance in ecosystems is owing to the increasing anthropogenic load on the environment. As such remedies can be natural physiologically active compounds, which include humic substances (HS) contained in all natural environments, including natural waters, soils, peat, sapropel, and coal. The HS formation is the second-scale (after the photosynthesis) process of transformation of organic substance in the nature, which involved about 20 GT of carbon per year. Humic substances perform a number of important ecological functions in the biosphere, such as accumulation, transport, regulatory, physiological and protective. Of particular relevance there becomes a study of the protective function of HS with a view to its further practical application, as it is responsible for the maintenance of equilibrium in ecosystems, which subjected to strong anthropogenic pressures.

The object of the research is the territory of operation of LLC "Gazprom dobycha Yamburg" on the Taz Peninsula, located in the North of Western Siberia in the Yamalo-Nenets Autonomous district (YANAO) between the Ob and Taz rivers. The natural ecosystems are southern and typical tundra with lichen, meadow, marsh, shrub and other vegetation with various crioturbated peat and podzol soils (Physico-geographical regionalization…1973). Ideal conditions for reindeer herding has Tazovskiy Peninsula, with a characteristic lichen land. However, the Taz Peninsula has a unique natural gas fields, which is a subsidiary company of Gazprom company - LLC "Gazprom dobycha Yamburg" conducting geological exploration, construction and development of new deposits, and also carries out the extraction of gas and gas condensate and their preparation for transport (Andreev et al. 2011). In the process of this kind of production activity are not excluded, in particular, the mechanical impact on soil and vegetation in the transportation equipment related to the implementation of exploration, drilling and construction of surface field facilities, when the tundra soils are partially or completely deprived of vegetation cover and organogenic layer, while mineral horizons appear on the surface and changes the cryogenic environment (Vasilevskaya & Kirillin 1993, Vasilevskaya & Grigor'ev 2002, Andreev et al. 2003). Meanwhile, LLC "Gazprom dobycha Yamburg," following the principles of the triune concept of sustainable development in all aspects of production activities is guided by the balance between the economic, social and environmental components (Andreev et al. 2011). One of the fundamental principles of the environmental policy of LLC "Gazprom dobycha Yamburg" is to minimize negative impacts and preserving the natural environment in the areas of placement of its production facilities, which is implemented in particular in the reclamation of disturbed tundra soils. For example, on soils devoid of organic layer creates a fertile layer (5-6 cm thick) by coating a mixture of peat and sand in the ratio 1:4, (Andreev et al. 2003). However, this method of remediation requires the use of large amounts of peat, which is a nonrenewable natural resource, at least within the technological cycle of gas production at this field. Therefore, it is necessary to seek methods of differentiation doses of peat depending on the physico-chemical properties of recultivated soils, as ways to enhance remediation parameters of peat. The latter may have been achieved through joint application of local peat and isolated from it humate (Bashkin 2016, Bashkin et al. 2016).

Accordingly, the aim of this work was the physico-chemical identification of humic substances, extracted from peat on site Yamburg gas condensate field.

Specific task phase of the study was theoretical substantiation and the application of NMR C13 spectroscopy to assess the purity of the preparations of humic substances isolated from peat and tundra ecosystems in the impact area of the Yamburg gas condensate field. To solve this problem 3 local peats were sampled from different peat quarries such as various natural depressions and excavations. The analytical work on the identification of humate structures of isolated compounds was conducted using modern analytical techniques: IR and NMR spectroscopy.

1. Theoretical Justification of the Method Using 13C-NMR Spectroscopy for Analysis of Humic Substances

Quantitative 13C NMR spectroscopy is one of the most informative methods of physico-chemical studies of the chemical nature of humic acids (HA) [6]. The applicability of this method in studies of the functional group, was confirmed repeatedly (Kalabin et al. 2000, Sartakov 2008, Chechenin et al. 1998).

In the analysis of most types of natural organic raw materials and its processed products (oil, coal, peat and other organic substances of the lithosphere, vegetable raw materials), it is impossible not to take into account their fundamental features, such as complexity, structural and phase heterogeneity, polydispersity, etc. These features determine not only the variation in a wide range of properties of these objects, but significantly limit the applicability of the standard chemical and physico-chemical methods of analysis, in particular, the most widely used atomic spectrometry, spectrophotometric, chromatographic and electrochemical ones. Quantitative presentation in terms of familiar structural formulas, i.e., in the form of component composition, is simply impossible. This is due to the fact that in addition to the variable composition and structure of the numerous components, the total structure and properties of the overall system is determined not only by covalent chemical bonds in individual molecules but weak non-valence interactions, such as compexation, hydrophobic interaction, hydrogen bonding, self-assembly etc. Thus, application of various destructive methods of analysis is inefficient. Among widely used non-destructive methods of analysis of complex organic mixtures of natural origin sufficient flexibility have such methods of spectroscopy as optical (IR and UV) spectroscopy, electron paramagnetic resonance (EPR) and nuclear magnetic resonance (NMR) spectroscopy (Kushnarev et al. 2013).

NMR spectroscopy on nuclei ^{1}H and ^{13}C today – one of the most informative methods for identification and structure determination of individual organic compounds of natural and synthetic origin and quantification of structural fragments of multicomponent mixtures of natural origin that are inaccessible to the analysis by other chemical and physico-chemical methods. The prospects of using NMR spectroscopy to study structural-group composition of soluble multicomponent systems are due to the followings:

- All the basic elements that are included to organic and natural substances and materials have stable isotopes, which NMR spectra can be registered (hydrogen - ^{1}H and ^{2}H, carbon – ^{13}C, etc.);
- Integrated intensity (area) of individual signals in the spectra is directly proportional to the relevant number responsible nuclei of groups of atoms, when using a correct experimental method of spectra registration;
- The same type of kernels belonging to different molecular fragments, may be characteristic for these fragments chemical shift (frequency of absorption), thereby enabling differentiated determination with the presentation of the results in the form of a fragment of the whole;
- The fragment analysis based on data set of NMR spectra is not required when the spectra using reference standards and internal standards, as based on the stoichiometric ratios of the content of the individual fragments relative to their total amount;
- Information on the molecular structure of objects of study can be represented at various levels of discretisation (slice, group structure, composition) or in the form of special physico-chemical, physiological and other characteristics;
- The results of analyses by the method of NMR spectroscopy can be adapted for calibration of other, more accessible physical and chemical methods.

The method of 13C NMR spectroscopy allows obtaining unique information about the dominant types of carbon–carbon bonds and their spatial arrangement. A large range of changes in shielding of the nuclei

13C (significantly greater than 200 m^{-1}), predetermines the low possibility of overlapping even broadened resonance signals of the carbon nuclei, make up the various structural fragments. There is a fairly complete classification of the chemical shifts of the nuclei of the carbon atoms on the types of relations (Brown & Ladner 1960, Brown et al. 1960), identifying the basic structural fragments of multicomponent mixtures of organic molecules.

The method of 13C NMR spectroscopy is informative enough for the identification and quantification of different oxygen-containing functional groups such as carbonyl (ketones and quinones), carboxyl (carboxylic acids and esters), alkoxylate (alcohols, phenols, ethers). A large number of works devoted to the study of the composition and structure of the products derived from oil, when liquefaction and obtaining soluble products from coal (extraction, thermal dissolution, hydrogenation) and humic substances. This is due to the greater efficiency of 13C NMR spectroscopy in comparison with 1H-NMR, the study of the structure of the compounds included in the composition of complex multicomponent systems of natural origin, including peats, coals, and their extracted different groups of compounds.

Currently, there is interest in the study of humic acids, in particular, by applying 13C NMR spectroscopy, which can be used to quantitatively study the structure of humic substances in general and humic acids (HA) in particular (Kovalevsky et al. 2000, Perminova 2000, Chukov 2001). The summarizing works are published on the theoretical foundations of spectroscopic methods and the interpretation of the spectra of organic compounds using 13C NMR spectroscopy (Silverstein et al. 2011).

Because of the complexity of the structure, the unique a wide range of interactions that can join HC and, in particular, their most reactive part, humic acids (HA). The presence of such groups as carboxyl, hydroxyl, carbonyl, combined with the presence of aromatic structures provides the ability of HS to join the ion and donor-acceptor interactions to form hydrogen bonds, actively participate in sorption processes. So, HC is well bind water, are capable of ion exchange, form complexes with metals and are adductive to various classes of organic compounds. Possessing specified properties, the HC performs a number of important biosphere functions. These include the structuring of the soil, the accumulation of nutrients and trace elements available to plants, the regulation of biogeochemical fluxes of metals in aquatic and soil ecosystems (Orlov 1990). By the end of the twentieth century, which one of the main problems is the chemical pollution of the environment, HC also get protective function. Under the HC protective action we mean their ability to help living organisms to withstand adverse conditions (stress factors) of the environment. So in the presence of chemical stressors (toxicants) HC can associate into stable complexes the metal ions and xenobiotics in contaminated water and soil environments (Orlov 1990). HC can be considered as natural detoxifiers. In other abiotic stress conditions (unfavorable temperature, pH, humidity, etc.) HC also have the ability to reduce the negative impact of different stressors on organisms both in aquatic and in soil environments.

A breakthrough in the understanding of the structure of humic substances has been the use of NMR spectroscopy (Wilson 1987). This method was first established that HC contains a higher percentage of lower aliphatic and phenolic structures than it is possible to determine the destructive methods. On the basis of the NMR data was first modeled the structure of marine humic substances as crosslinked aliphatic chains having at its ends a carboxyl group. By using NMR for the first time in HC were also found acyl oligosaccharides and terpenoids (Leenheer et al. 2003, Leenheer 2007). The undeniable advantage of NMR spectroscopy is that it allows to obtaining quantitative data on the content of HC of different structural fragments (Kovalevsky et al. 2000). In addition, the two-dimensional NMR experiments allowed to establishing specific connections between atoms of carbon, hydrogen and nitrogen in HC.

When determining the structural-group composition by 13C NMR spectroscopy as the main problem considers the establishment of a set of structural fragments that fully describe the distribution of the main structure-forming element, carbon, in the HC structure, the content of which was amenable to quantitative definition. The best method for solving this problem is one-dimensional NMR spectroscopy on 13C nuclei. Typical 13C NMR spectra of different origin and fractional composition HC are shown in Figure 1.

13C NMR spectra are characterized by broad absorption bands due to overlapping of a large number of signals. Maximum intensity is achieved in the field unsubstituted aliphatic fragments (HC), "carbohydrates" (HCO fragments) and aromatic (Ar) region of the spectrum. Particularly intense, strongly overlapping peaks are observed in the "carbohydrates" area. This distribution of the spectral density is consistent with a priori information about the existance in structure of humic compounds the highly replaced aromatic skeleton and a carbohydrate-peptide periphery. Because of the strong overlapping, the assignment of signals in these spectra is only possible by "interval" in accordance with the position of the resonances of atoms with similar chemical environment, as shown in Table 1 (Kulikova 2008).

Table 1. The main lines in the 13C NMR spectra of humic substances

Fragment	Interval, m^{-1}	Description
CH-H,C	5-48	H,C – replaced aliphatic fragments
Alpha-C,H		Aliphatic fragments in alpha-position to electronegative group or aromatic ring
CH_3-O	48-58	Metoxilic and O-, N-replaced aliphatic fragments "carbohydrate"
CH_2-O,N	58-64	
CH-O,N	64-90	
O-CH-O,N	90-108	Aliphatic fragments twice replaced by heteroatoms – "anomeric"
Ar-C,H	108-145	H-, C-replaced aromatic fragments
Ar-O,N	145-165	O-, N-replaced aromatic fragments
COO-H,C	165-187	Carboxil groups and their derivates
C = O	187-220	Ketone and quinone groups

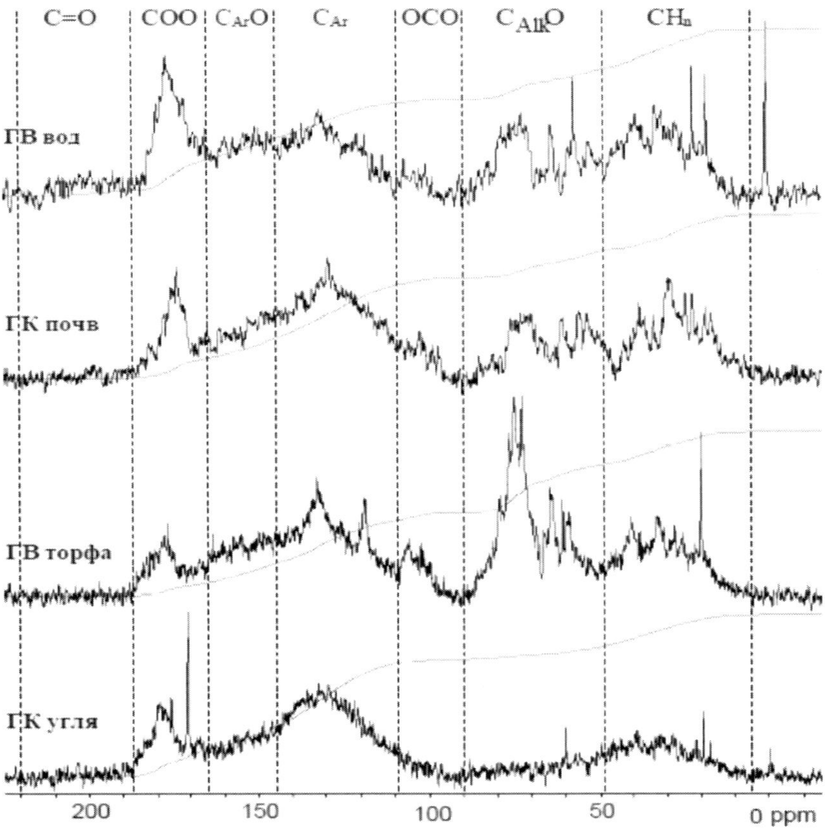

Figure 1. Typical 13C NMR spectra of different origin and fractional composition HC: up from the bottom – coal, peat, soil, water.

The maximum level of detail allows you to highlight in the 13C NMR spectrum of 9 intervals, integration of which can be carried out with precision. The magnitude of these partial integrals, the sum of which describes the complete integral of the spectrum, can be used as integrated descriptors of the structural-group composition, if they are proportional to concentration of atoms in the relevant parts of fragments (Perminova 2000). Due to the difference of relaxation times of nuclei of different types, the proportionality of the magnitude of the integral number of atoms in the 13C NMR spectrum is achieved only in quantitative terms of its registration (Kovalevsky et al. 2000). They include (1) using time delay between pulses (Td), sufficient for complete relaxation of the tertiary and quaternary carbon nuclei, and (2) elimination of the nuclear overhauser effect.

The carbon content of various types were determined by integration of the corresponding spectral regions (m^{-1}): 220-187 – ketone and chinone groups (CC = O); 187-165 - carboxyl and ester groups (CCOO); 165-108 - aromatic fragments (CAr); 108-5 – aliphatic fragments (CAlk).

Thus, quantitative 13C NMR spectroscopy is one of the most informative methods for physico-chemical studies of the chemical nature of humic acids (HA). The applicability of this method in studies of the functional group was confirmed repeatedly.

2. C13 NMR Spectroscopy for Analyzing Humic Substances Extracted from Peat of Yamburg Gas Condensate Field

2.1. Extraction of Humates from Peat Samples from the Territory Yamburg GCF

Humus refers to all organic compounds in the soil, except the part of living organisms and organic residues preserved anatomical structure. In humus composition non-specific organic compounds and specific humic substances are subdivided. Nonspecific is a group of organic substance known to nature and the individual constitution. They enter the soil from decaying plant and animal residues and root secretions. Nonspecific compounds represented by almost all the components that make up animal and vegetable tissues and in vivo selection of macro — and microorganisms. These include lignin, cellulose, proteins, amino acids, sugars, waxes and fatty acids.

Overall, the proportion of non-specific organic compounds does not exceed 20% of the total amount of soil humus. Nonspecific organic compounds are products of different degrees of decomposition and humification entering the soil from plant, animal and microbial material.

These couplings determine the dynamics of a rapidly changing soil properties: the redox potential, the content of mobile forms of nutrients, the abundance and activity of soil microorganisms, composition of soil solutions. Humic substances, on the contrary, contribute to the stability over time of certain properties of the soil: exchange capacity, water-physical properties, air regime and color.

A variety of natural humic acids (HA) is large, and accordingly their extraction and identification are depending from the methods used. The only real symptom that is used in practice is the ability of HA to dissolve in alkalis and to form precipitation upon acidification of the solution. But today there is a great variety of methods for the isolation of humic acids, the most common methods of extraction by solutions of alkalis and alkaline salts, less well – known methods of extraction with organic solvents (Orlov 1990).

The last group of methods is based on the use of organic solvents – acetyl bromide, aqueous dioxane, furfural, amines and amino alcohols of the fatty series. They have been little studied and are based only on the assumption that all organic matter in the solid peat, with the exception of HA, soluble in the acetyl bromide. On the contrary, in water-dioxane, furfural and other solvents only HA is soluble, while the other constituent parts of peat remain undissolved (D'yachkov 1979).

It is assumed that HA extracted by these methods are cleaner and less altered than in the extration of alkaline solutions. However, the study of the relationship of the total HA to that dissolved in listed solvents was showed that they are only partially soluble in dioxane, acetone. In these solvents dissolves a part of the rest of the organic mass of peat. Therefore, the use of organic solvents may not be recommended for the extraction of HA from peat.

The most common methods are based on the dissolution of humic acids in aqueous solutions of alkalis and alkaline salts with soluble humates formation and acidification of the solution to precipitate the free humic acids (D'yachkov 1979, Physics and chemistry...1989, Mineev 1989, Maryganova et al. 2003, Bambalov & Belenkaya 1998, Yudina et al. 1996, Aleksandrova 1980, Martynova 2011, Mineev 2001, Orlov &

Grishina 1981, Dementieva 2011, Lishtvan 1975). To do this, use the hydroxides of sodium and potassium, ammonia, soda, sodium fluoride, acetate, oxalate and sodium pyrophosphate, ammonium oxalate. Moreover, for the extraction of humic acids sodium hydroxide can be used in various concentrations: 0.08% or 0.02 n; 0.4 percent or 0.1 n; 0,5%; 0,8%; 1%; 2%; 4-10%. Accordingly the number of alkali treatments may be: 4-10; 3-4; 2; 1. The extraction can be carried out at 20°, 80°C and 100°C. Use of a boiling alkaline solvent promotes the extraction of humic acids, which lose its solubility when dried, furthermore enhances the solubility of bitumen and pectin. Drying of selected humic acids, prone to oxidation and decomposition, is carried out at a temperature of 30-40°, 60°, 70°, 80°, 105°C and even at 140°C.

2.2. Conducting Analytical Work to Identify the Structures of Isolated Humatic Compounds Using Modern Analytical Methods

Humic acid (HA) of peat is a complex mixture of high molecular weight and polyfunctional alicyclic compounds, hydroaromatic, aromatic and heterocyclic nature, replaced with different length alkyl chains both normal and isomeric structure, including saturated and unsaturated, with the various functional groups (carboxyl, hydroxyl, metaxylene, amide, and amino groups) (Orlov 1990). Humic acids of peat represent an acid-base complex are donors and acceptors of the anions, cations, protons and electrons. These features determine not only the great diversity of chemical properties of these acids, but a wide range of their biological activity. Thus, despite the presence of a significant number of common elements of the structure, the biological activity of HA obtained from different deposits of peat can vary considerably because of the high degree of polymorphism in their chemical structure. This makes virtually impossible attempts of theoretical generalization and prediction of YA structure and composition. Accordingly, it requires the experimental characteristics of the properties on the basic elements of the structure and necessitates the biological standardization of samples from each peat field. Thus we carried out the analytical studing the chemical structure and properties by using NMR and IR spectroscopy of humic acids obtained from 3 peat samples from the territory of Yamburg GCF.

Specific organic part of the soil humic substances is a heterogeneous (heterogeneous) system of polydisperse high-molecular nitrogen-containing aromatic compounds of acid nature. Humic substances are formed as a result of complex biophysico-chemical process of transformation (humification) of products of decomposition of organic residues entering the soil. Depending on the chemical composition of plant residues and factors of decomposition (temperature, humidity, composition of micro-organisms) there are two main types of humification: fulvates and humates. Each of them corresponds to a certain fractional-group composition of humus. The most important groups are humic acid (HA) and fulvic acids (FA).

Humic acids contain a 46 — 62% of carbon (C), 3 to 6% nitrogen (N), 3-5% hydrogen (H) and 32-38% of oxygen (O). In the composition of fulvic acids there are lesser carbon content — 45-50%, nitrogen — 3,0—4,5% and of hydrogen is 3-5%. Humic and fulvic acids almost always contain sulfur (up 1.2%), phosphorus (tens to hundreds of a percent) and cations of different metals.

2.3. Extraction of Humates

Extraction of humates from 3 peat samples was performed in accordance with the following method:

1. Preliminary decalciumization of sample by 0.1 n H_2SO_4 solution. The suspension was left for a day. After settling, the solution from the solid phase was separated by decantation.
2. Extraction of humic substances was carried out in 0.1 n NaOH at room temperature. The relationship between hitch and extractant is 1:15. The suspension was stirred periodically and left overnight.

The extraction duration is 20 hours. Solid phase from the alkaline solution was separated by centrifugation. The subsequent 4-5 extractions are necessary. The alkaline extracts were combined, then the solution of humates was centrifuged to separate the colloidal particles.

3. For precipitation of HA 10% HCl is used, i.e., ~ 20 ml per liter of extract to establish a pH of 1-2. Left for a day to the full deposition. HA precipitate was separated by centrifugation.
4. Cleansing preparations of HA by re-deposition. HA precipitate is re-dissolved in 0.5-1.0 l of 0.1 n NaOH. Sodium sulfate was added for coagulation of finely dispersed mineral particles. Alkaline HA solution was centrifuged for separating solids. Then HA was precipitated by adding 0.1 n HCl to establish a pH of 1-2.
5. HA precipitate is repeatedly washed with distilled water to pH~6 by universal indicator paper.
6. Drying the precipitate of HA in the cupboard at 50°C.

The relative outputs of HA preparations from 3 peat samples from the studied territory of Yamburg GCF were: peat sample No 8 - 5.78%, No 9 - 17,75% and No 10 - 16.68%.

For long-term storage, as well as for the experimental studies with the aim of analyzing the structure of humic substances by spectral methods (IR and NMR) were used the resulting dry preparations. Definition of group and fractional composition of humus is based on Tyurin scheme in modification Ponomareva and Plotnikova (Mineev 2001, Orlov and Grishina 1981).

The results on the fractional composition of the humus of the three peats from the studied territory are presented in Tables 2-5.

Group composition of humus characterizes the proportion of humic acids and fulvic acids. A quantitative measure of the type of humus is the ratio of the carbon content of humic acids (HA) to the carbon content of fulvic acid (FA). The biggest difference among the 3 studied peats is observed in the total carbon content and FA fractions (Tables 4-5). These data should be considered when conducting experiments aimed at the selection of peat with maximum biological activity of various peat bogs for the purpose of reclamation of disturbed and contaminated soils, including stimulation of the growth and development of vegetation on reclaimed disturbed and contaminated soils with the help of special preparations, including extracts from samples of peat under controlled hydrothermal conditions with the aim of developing solutions for optimizing soil remediation.

Table 2. The content of total carbon in peat samples of Yamburg GKM

Peat sample	Carbon total content,%
8	14.4 ± 0.3
9	34.9 ± 0.4
10	27.4 ± 0.9

Table 3. The content of organic carbon in the extracts from peat (C, % 100 g peat)

Sample	Direct extraction by 0.1 n NaOH	Extraction by 0.1 n H_2SO_4	Extration by 0.1 n NaOH after decalciumization
8	4.9 ± 0.1	0.13 ± 0.03	5.2 ± 0.1
9	10.5 ± 0.1	0.33 ± 0.01	11.1 ± 0.1
10	8.9 ± 0.1	0.32 ± 0.02	9.1 ± 0.1

Table 4. Fractional composition of humic substances, in% per 100 g of peat

Sample	HA, %			FA, %				HA +FA, %	HA/FA
	1	2	sum	1a	1	2	sum		
8	3.16	0.07	3.23	0.13	1.61	0.36	2.10	5.3	1.5
9	6.86	0.53	7.39	0.33	3.31	0.40	4.04	11.4	1.8
10	5.58	0.49	6.05	0.32	3.00	0.05	3.37	9.4	1.8

Table 6. Fractional composition of humic substances, % from Ctot

Sample	HA, %			FA, %				HA+ FA, %	HA/FA
	1	2	sum	1a	1	2	sum		
8	21.9	0.5	22.4	0.9	11.2	2.5	14.6	37.0	1.5
9	19.7	1.5	21.2	0.9	9.5	1.1	11.5	32.7	1.8
10	20.4	1.8	22.2	1.2	10.9	0.2	12.3	34.5	1.8

Note:
HA fractions:
fraction 1 – free and bound with mobile sesquioxides oxides;
fraction 2 – bound calcium
FA fractions:
fraction 1A – free and bound with mobile sesquioxides oxides;
fraction 1 is associated with the 1 fraction of humic acids;
fraction 2 – bound fraction 2 humic acids

Functional composition of humic substances from peat was studied using the method of IR-spectroscopy. This method allows obtaining information not only about the set of critical atomic groups, and types of relationships, but also about the specific location of individual groups. Identification of IR spectra for humic substances was carried out with involvement of chemical and physical methods as well as literature data. The use of relevant identification tables of the IR spectra (Kovalevsky et al. 2000) has allowed interpreting functional parts of peat humic acids. The measurements were performed using IR-Fourier spectrometer Nicolet 6700 XT by Thermo Electron Corporation (USA) in tablets with KBr in the ratio 1:200, respectively, in the range of frequencies from 400 cm^{-1} to 4000 cm^{-1}.

The spectra of infrared absorption (Figure 2) of studied HA indicate typical absorption bands [16]. It has a broad intense absorption band with a maximum at 3500-3300 cm–1, which is caused by valence vibrations of OH groups (νOH) of different nature, predominantly hydrogen bonds. In the longwave wing of the main line about 3 250-3, 200 cm^{-1} one revealed the absorption of medium intensity, having the form of a shoulder (inflection) and meet the stretching vibrations N—H (νNH) in the structure of the amide and amine linked by hydrogen bonds. At the values 2 926 - 2 920 and 2 852 cm^{-1} there are clearly visible the bands of medium intensity, absorbing due to the stretching vibrations of $–CH_3$ and $–CH_2$-groups of the side chains in HA molecules, mainly due to methylene groups, as indicated by the asymmetric band and the symmetric deformation vibration in the region 1 480— 1 380 cm^{-1}.

On the spectra of the studied HA one observed weak absorption at 2 600-2 500 cm^{-1}, which is typical for carboxylic acids. This broad band is considered to be very characteristic for stretching vibrations of hydroxyl groups involved in the formation of strong hydrogen bonds in the dimeric forms of the carboxylic acids. A distinct maximum of the strong intensity found in the interval from 1 725 to 1 650 cm^{-1}, which is characteristic for stretching vibrations of carbonyl group (νC = O), which can be represented by ketones, aldehydes, carboxylic acids and their functional derivatives (Orlov 1990). An intense absorption band in region of 1 650 - 1 600 cm^{-1} is due to in-plane vibrations of conjugated carbon-carbon (aromatic, νC = C) and carbon-oxygen bonds (carbonyls associated by hydrogen bonds, carboxylation, νC = O), aromatic skeletal and quinones. Some authors believe that it is the aromatic νC=responsible for this band, the maximum of which falls on the 1610 cm^{-1}. In the study of HA maximum is observed also at 1 630 cm^{-1}, which may be due to the superposition of bands of amide groups (δNH_2, amide II). The absorption band around 1 513 cm^{-1} (νC—C) indicates the presence of uncondensated aromatic compounds in the HA composition associated with atoms of nitrogen and oxygen.

Absorption in the 1264-1225 cm-1 is mainly defined by valence (νC—O) and deformation vibrations of (δ O—H) bonds. The absorption of radiation in the shortwave part of the spectrum in the region 1150-1000 cm^{-1} stretching vibrations responsible for alcohol and carbohydrate hydroxyl groups (νOH). In the range of about 1050 cm^{-1} (νC—O) radiation is absorbed by primary alcohols, at 1100 cm^{-1} (νC—O) — secondary, and at 1 150 cm^{-1} (νC—O) — tertiary alcohols. The absorption in this area can also be caused by valent oscillations (νC—O—C) of glycosidic linkages of carbohydrates, C—O—S-bonds of cyclic and aliphatic ethers. In the region of wave numbers from 1000 to 600 cm^{-1} weak absorption bands is observed, possibly due to off-plane deformation vibrations (δC—N) in the aromatic rings with two or more unsubstituted hydrogen atoms.

Thus, the maximum intensity of the absorption bands in the spectra of HA of peat is observed for hydroxyl, carbonyl (carboxyl) groups, aliphatic and aromatic fragments.

Quantitative estimation of contents of functional groups are given on the basis of the relationship of optical density of absorption bands of oxygen-containing groups to the optical densities corresponding to polyisoprenyl aromatic systems and aliphatic substitutes at 2920 cm^{-1}. Most informative in this case are the ratio of GK hydrophobic and hydrophilic components.

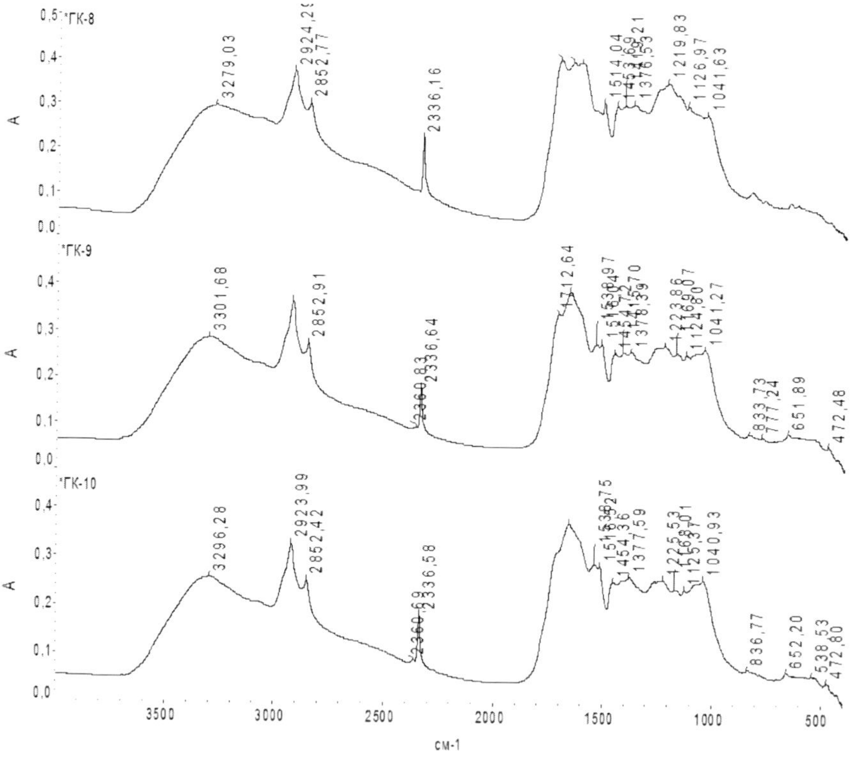

Figure 2. The IR spectra of humic substances from HA of peat 8 (uppur spectra), 9 – middle spectra and 10 - lower spectra.

Table 6. The ratio of the optical densities of the absorption bands at certain wavelengths in the humic acids of peat according to IR-spectroscopy

HA, peat	3300/1610	2920/1610	2860/1610	1710/1610	1260/1610	1650/1610	3300/2920	1710/2920	1260/2920
8	0,75	0,95	0,75	1,02	0,85	1,01	0,79	1,08	0,90
9	0,84	1,08	0,80	1,01	0,79	1,15	0,78	0,93	0,74
10	0,76	0,96	0,74	0,93	0,75	1,10	0,79	0,97	0,78

Quantitative estimation of contents of functional groups are given on the basis of the relationship of optical density of absorption bands of hydroxyl groups D_{3300}, $-CH_3$, $-CH_2-$ groups D_{2920}, D_{2860}, carbonyl-containing groups D_{1710}, hydroxyl groups in alcohols, carbohydrates D_{1260} to an optical density corresponding to the aromatic polyisoprenyl systems D_{1610}. The calculations of the structural parameters HA are given in Table 6.

One of the main forms of oxygen in HA of peat are hydroxyl, carboxyl groups, C-O–bond at 1260 cm^{-1} and –CO -OH of hydrocarbons. Their relative content depends on the degree of transformation of humic substances. The ratio of optical densities of absorption bands of oxygen-containing functional groups and alkyl substituents to aromatic fragments showed the dominance of the first over of the alkyl (2920 cm^{-1}) and C-O bonds (1260 cm^{-1}). Peat 8 is different in this index from peat 9 and 10.

From the Table 6 data it follows that the molecules of peat HA carboxyl groups are dominated on alkyl substituents, the ratio D_{1710}/D_{2920} for all samples of peats more than 1. The close values of relations D_{1710}/D_{2920} for all the HA samples characterize them as structures with a similar system of polyisoprene system of H-bonds.

The relative number of hydroxyl groups (D_{3400}/D_{1610}) in HA of peat is not high. The value of these indicators for the most of the peat HA ranges from 0,75-0,84. The proportion of carboxyl groups in HA of peat, (D_{1710}/D_{1610}) varies between 0,93-1,02. There is a change in the relative content of carboxylic groups in HA from peat 10 to 8; that may be due to differences in their botanical composition. In favor of this view one shows also the difference of the relative content of aliphatic linkages relative to the aromatic (D_{2920}/D_{1610}). This value is, presumably, determined the botanical composition of peat.

The results show that the number of oxygen-containing groups of all types below the number of aliphatic C-H bonds. For peat HA the same value is observed for the spectral coefficient D_{3300}/D_{2920}, obviously, this is due to a single occurrence of oxidative processes in the upper layers, in the zone of fluctuation of groundwater

Table 7. Quantitative comparison of the 13C NMR spectra of peats from the territory of Yamburg GCF

0–50 ppm Aliphatic structures (%)
Peat 8 37.9
Peat 9 48.5
Peat 10 54.03
50–100 ppm O-alkyles (%)
Peat 8 21.2
Peat 9 21.3
Peat 10 20.3
100–110 ppm Acetyl groups (%)
Peat 8 5.57
Peat 9 3.23
Peat 10 2.76
110–160 ppm Aromatic structures (%)
Peat 8 23,22
Peat 9 16,1
Peat 10 14,09
160–190 ppm Carboxyl groups (%)
Peat 8 10.07
Peat 9 9.26
Peat 10 7.94
190–230 ppm Carbonyl groups (%)
Peat 8 1.99
Peat 9 1.61
Peat 10 0.91

As a result of elemental analysis and IR-spectroscopy it was shown that the studied HA represent the compounds with a high share of aliphatic fragments, represented by long conjugated chains. Hydrolyzed (peptides and carbohydrates) compounds are predominant in the composition of the peripheral aliphatic structures, the contribution of non-hydrolyzed compounds (S $-CH_3$ and $-CH_2$ - groups) in the aliphatic part is insignificant. Humic acids have a low degree of benzoindole and contain active oxygen and nitrogen containing functional groups (carboxyl, hydroxyl, amide, amino groups).

With the aim of obtaining information about the content of structural fragments in the peat HA the method of 13C NMR spectroscopy was used. Registration of spectra was performed with the radio spectrometer BRUKER AVANCE-II NMR 400 WB at a frequency of 100.4 MHz using the technique of Fourier-accumulated transformation, which allowed to determining the relative content of magnetic carbon nuclei belonging to a particular group of atoms.

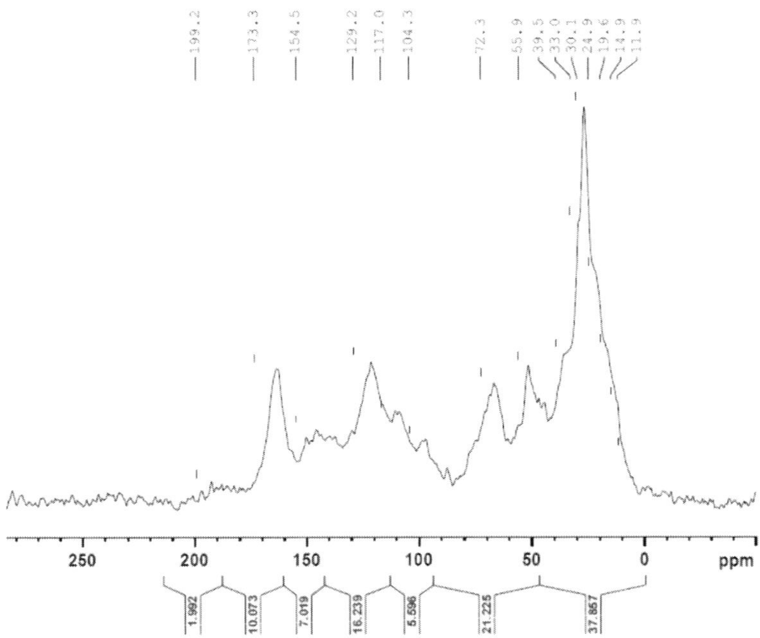

Figure 3. The 13C NMR spectrum, peat 8.

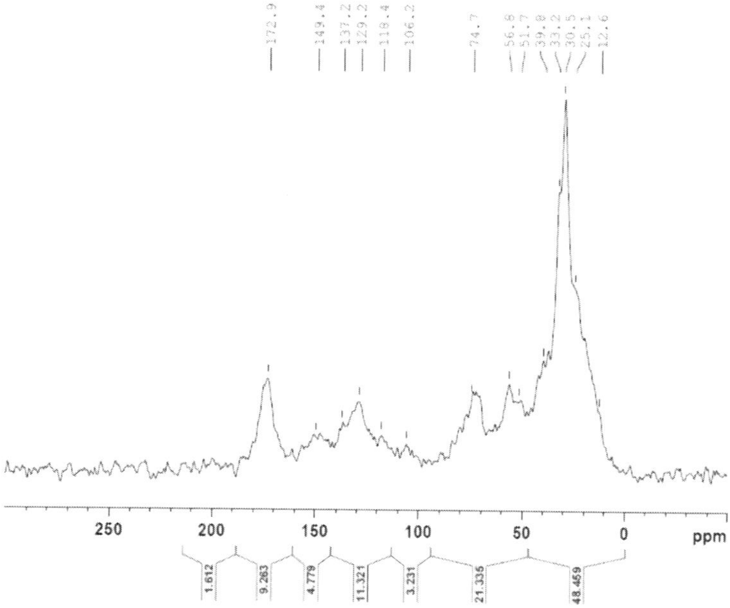

Figure 4. The 13C NMR spectrum, peat 9.

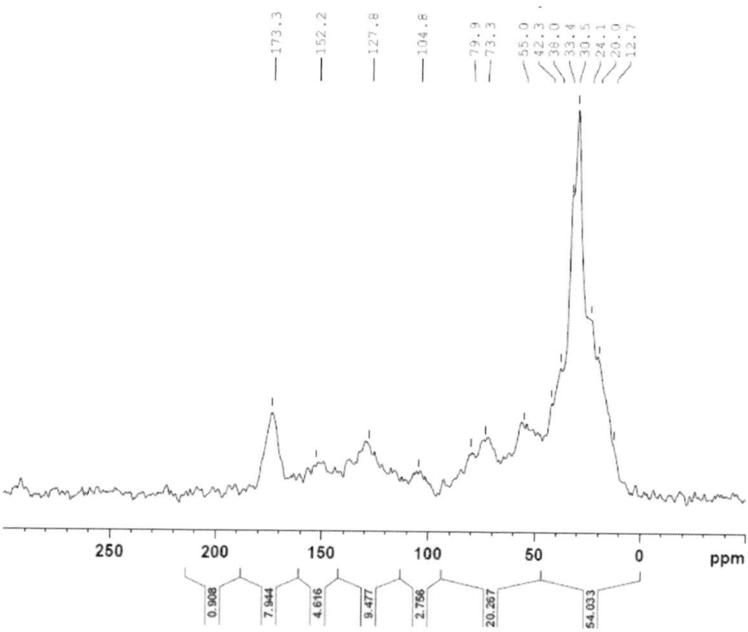

Figure 5. The 13C NMR spectrum, peat 10.

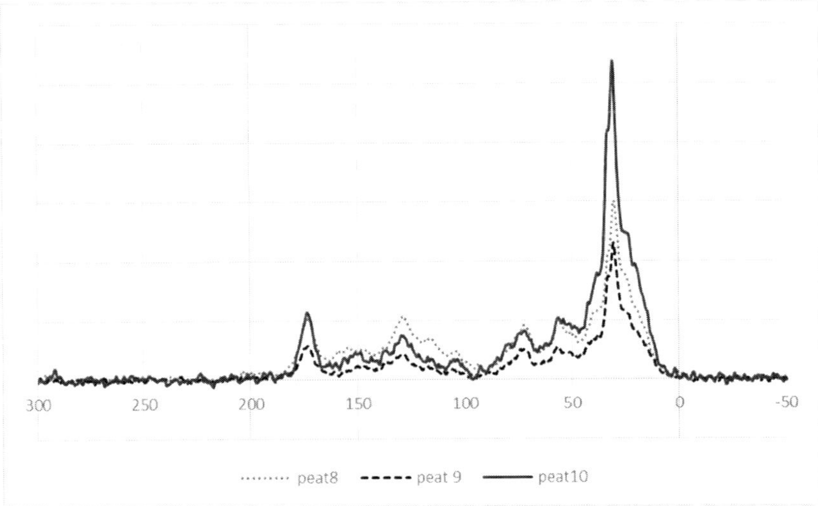

Figure 6. Graphical comparison of the 13C NMR spectra of peats from the territory of Yamburg GCF.

Spectra were shooting with a delay of 2 s for CP/MAS between pulses. 13C chemical shifts were calibrated relative to the CH_2 groups (38.48 ppm) of solid adamantane ($C_{11}H_{15}ClO$), which was used as external standard. Using 13C NMR spectroscopy the information was obtained on the distribution of carbon atoms in different structural groups in macromolecules of peat HA (Table 7; Figures 3-6). Analysis of 13C-NMR spectroscopy showed the presence in humates of aliphatic and aromatic carbon atoms associated with oxygen and represent a variety of functional groups – hydroxyl, carboxyl, phenolic, alcoholic, and carbohydrate.

In the spectra the 13C-NMR there are identified the following intervals corresponding to resonant absorption of carbon nuclei: 0-50 ppm – saturated unsubstituted carbon atoms, 54-66 ppm – carbon atoms in metaxylene groups, 66-106 ppm – oxidized aliphatic fragments, 110-140 ppm and aromatic carbon atoms, 140-160 ppm – oxidized aromatic fragments, 160-200 ppm – carbon atoms in carbonyl-containing groups. It is possible to identify the following fragments of the macromolecule, HA: alkyl (aliphatic) carbon (0-50 ppm), called for short polysaccharide carbon in carbohydrates, alcohols, acetals (50-110 ppm), aromatic carbon (110 - 160 ppm), carboxyl C (160-200 ppm).

A common feature of peat HA is the high proportion of aliphatic carbon. In the aromatic structures it contains up to 23 wt.% carbon and from 38 wt.% to 54% in saturated aliphatic fragments. As follows from previous studies of peat in the region (Sartakov 2008), the fragment composition of peat is determined by the characteristics of the original plants of the peat-forming plants and the conditions of their transformation. In contrast to the peat (Sartakov 2008), we observed a higher content of aliphatic carbon upto 54% (peat 10) and a smaller proportion (about 2-told) of aromatic carbon upto 14% (peat 10). In a number of studied peats the following regularity is observed: with the consistent increase in the percentage of C (aliphatic) and decreasing the content of C (aromatic) by peat peat 8 to 10, which is the rate of changing the degree of peat decomposition and may also depend on its botanical species.

CONCLUSION

Quantitative 13C NMR spectroscopy is one of the most informative methods physico-chemical studies of the chemical nature of humic acids (HA). The applicability of this method in studies of the functional group were repeatedly confirmed.

As a result of our analysis of methods for the preparation of humates from peat, we extracted three fractions of humic acids and four fractions of fulvic acids, which differ in structure and characterizing features of the humic complex of peat.

Group composition of humus characterizes the proportion of humic acids and fulvic acids. The biggest difference among the 3 peats is observed in the total carbon content and fractions of fractions of fulvic acids, FA. As a result of elemental analysis and IR-spectroscopy it was shown that the studied HA represent a connection with a high share of aliphatic fragments, represented by long conjugated chains. Spectral methods capture some of the difference in humates derived from peat. These data should be considered when conducting experiments aimed at the selection of peat with maximum biological activity of various peat bogs for the purpose of reclamation of disturbed and contaminated soils, including stimulation of the growth and development of vegetation on reclaimed disturbed and contaminated soils with the help of special preparations, including extracts from samples of peat under controlled hydrothermal conditions with the aim of developing solutions for optimizing soil remediation.

The results of NMR spectroscopy of humic substances of peat from the territory of Yamburg GKM showed that a common feature of HA of peat is the high proportion of aliphatic carbon. The aromatic structures contain up to 23 wt. % carbon, and saturated aliphatic fragments - from 38 wt. % to 54%. As follows from previous studies of peat in the region, the fragment composition of peat is determined by the characteristics of the original plants of the peat-forming plants and the conditions of their transformation. In contrast to the earlier study of peat HA in more southern tundra region (Sartakov 2008), we observed a higher content of aliphatic carbon to 54% (peat 10) and a smaller proportion (about 2-told) of aromatic carbon to 14% (peat 10). In a number of studied peats the following regularity is observed: with the consistent increase in the percentage of C (aliphatic) and decreasing the content of C (aromatic) by peat 8 to peat 10, which is the rate of changing the degree of decomposition of peat and may also depend on its botanical species.

Conflict of Interest

The authors confirm that this article content has no conflicts of interest.

Acknowledgments

The authors express their deep appreciation and gratitude to their affiliation institutions for the support during research and development.

References

Aleksandrova, LN (1980) Soil Organic Matter and Processes of its Transformation. Nauka, Leningrad.

Andreev, OP, Stavkin, GP, Levinson, IL, Perepyolkin, BI, & Lobastov, SA (2003) Protecting and restoring the lands and landscapes of the far North during gas production. *Ecology and Industry of Russia*, 6, 4-9.

Andreev, OP, Bashkin, VN, Galiulin, RV, Arabsky, KA & Maklyuk, OV (2011) Problems of geoecological risks in gas industry. Overview. Gazprom VNIIGAZ, Moscow.

Bambalov, NN & Belenkaya, TY (1998) Fractional-group composition of organic matter of virgin and reclaimed peat soils. *Soil Science*, 12, 1431–1437.

Bashkin, VN (2016) Biogeochemical technologies for managing pollution in polar ecosystems. Springer, Switzerland.

Bashkin, VN, Galiulin, RV, Alekseev, AO, Galiulina, RA & Maltseva, AN (2016) Biogeochemical technology for Disturbed Tundra Soils Remediation by Peat Potassium Humate Application. In: Bashkin, V (Ed). Biogeochemical technologies for managing pollution in polar ecosystems. Springer, Switzerland, 211-218.

Brown, JK & Ladner, WR (1960) A study of hydrogen distraction in coal-like materials by high-resolution NMR spectroscopy. I. The measurement and interpretation of the spectra. *Fuel*, 39(1), 79-86.

Brown, JK, Ladner, WR, & Shepard, NA (1960) A study of hydrogen distraction in coal-like materials by high - resolution NMR spectroscopy. II. A comparison with infra-red measurement and the conversion to carbon structure. *Fuel*, 39(1), 87-96.

Chechenin, IE, Projdakov, AG, Paramonova, TG, & Kushnarev, DF (1998) Analysis of humic acids of Kondinsky field by the method of NMR spectroscopy. *Chemistry of Solid Fuel*, 2, 40-47.

Chukov, SN (2001) Structural-functional parameters of soil organic matter under anthropogenic impact. St-Petersburg State University, St-Petersburg.

Dementieva, TV (2011) Physicochemistry and biology of peat. a guide to the methods of studying the transformation of organic substances of peat. Tomsk State University, Tomsk.

D'yachkov, GS (1979). Humic acid of peat and methods for their determination. Solid Fuel Chemistry, 2, 130-135.

Kalabin, GA, Kanitskaya, LV, and Kushnarev, DF (2000). Quantitative NMR spectroscopy of natural organic raw materials and products of its processing. Chemistry, Moscow.

Kovalevsky, DV, Perminov, AB, Perminova, IV & Petrosyan, VS (2000) Choice of conditions of registration for quantitative 13C NMR spectra of humic acids. *Herald of Moscow State University, Ser. 2. Chemistry*, 41(1), 39-42.

Kulikova, NA (2008). Protective acting of humic substances in relation to plants in aquatic and soil environments in conditions of abiotic stress. Dissertation of Doctor of Biological Chemistry. Moscow State University, Moscow.

Kushnarev, DF, Projdakov, AG, and Bisikalo, AL (2013). Quantitative NMR spectroscopy of multicomponent systems of natural origin. Irkutsk State University, Irkutsk.

Leenheer, JA, Nanny, MA & McIntyre, C (2003) Terpenoids as major precursors of dissolved organic matter in landfill leachates, surface water, and groundwater. *Environ. Sci. Technol*, 37, 2323-2331.

Leenheer, JA (2007) Progression from model structures to molecular structures of natural organic matter components. *Annals of Environ. Sci*, 1, 57-68.

Lishtvan, II (1975) Basic properties of peat and methods for their determination. Nauka and Tekhnika, Mins.

Martynova, N (2011) Chemistry of soils: soil organic matter. Irkutsk State University, Irkutsk.

Maryganova, VV, Bambalov, NN, & Parmon, SV (2003) Influence of the type of extractant on the structure of extracted from a peat humic acids. *Chemistry of Solid Fuel*, 1, 3-10.

Mineev, VG (Ed) (1989) Workshop on agricultural chemistry, Moscow State University, Moscow.

Mineev, VG (ed) (2001) Workshop on agricultural chemistry, Moscow State University, Moscow.

Orlov, DS (1990) Soil Humic acids and general theory of humification. Moscow State University, Moscow.

Orlov, DS & Grishina, LA (Eds) (1981) Workshop on humus chemistry. Moscow State University, Moscow.

Perminova, IV (2000) Analysis, Classification and Prediction of Properties of Humic Acids. Dissertation of Doctor of Chemistry. Moscow State University, Moscow.

Physico-geographical regionalization of the Tyumen region (1973) Moscow State University, Moscow.

Physics and Chemistry of Peat. Moscow: Nedra, 304 pp.

Sartakov, MP (2008) 13C NMR Spectroscopy of humic acids of peats of middle Ob region. *Chemistry of Vegetable Raw Materials*, 3, 135-139.

Silverstein, R, Webster, F & Kimel, D (2011) Spectrometric identification of organic compounds. Binom, Moscow.

Vasilevskaya, VD & Kirillin, VV (1993) Anthropogenic soil degradation in the southern tundra of Yamal and the measures for their prevention. *Bulletin of Moscow State University, Series 17. Soil Science*, 4, 3-9.

Vasilevskaya, VD & Grigor'ev, VY (2002) Biological indicators of degradation and self-restoration the soil-vegetation cover of tundra. *Siberian Ecological Journal*, IX (3), 355-370.

Wilson, MA (1987) NMR techniques and applications in geochemistry and soil chemistry. Pergamon, New York.

Yudina, NS, Pisarev, SI & Saratikov, AS (1996) Evaluation of biological activity of peat humic acids. *Solid Fuel Chemistry*, 5, 31-34.

In: Ecological and Biogeochemical Cycling … ISBN: 978-1-53612-081-3
Editor: Vladimir N. Bashkin

Chapter 8

NEW ECOLOGICALLY ORIENTED BIOGEOCHEMICAL TECHNOLOGIES FOR RECLAMATING DAMAGED GAS PRODUCTION AREAS

Vladimir N. Bashkin[1,2], Rauf V. Galiulin[3] and Andrey O. Alekseev[1]

[1]Institute of Physicochemical and Biological Problems of Soil Science of Russian Academy of Sciences, Pushchino, Russian Federation
[2]Gazprom VNIIGAZ LLC, Razvilka, Russian Federation
[3]Institute of Basic Biological Problems of Russian Academy of Sciences, Pushchino, Russian Federation

ABSTRACT

Three new ecologically oriented biogeochemical technologies of reclamation of mechanically disturbed tundra soils of the territories of productive activity of the gas industry in conditions of the Far North are presented.

In the first technology the local peat dose depending on a soil full moisture capacity is defined, in the second technology the procedure of receiving from local peat a potassium humate as a stimulator of growth and development of plants is described, and the third technology is devoted to the combined use of local peat and potassium humate at soil reclamation.

Keywords: Far North, gas industry, mechanically disturbed soil, reclamation technology, local peat, dose, full moisture capacity, potassium humate

INTRODUCTION

At present, the quantitative parameterization of multi-scale local, regional and global changes due to natural and anthropogenic impacts requires the quantitative understanding of biogeochemical cycles and it seems to be one of the fundamental branches of modern science.

In turn, studying the fundamental mechanisms of quantitative parameterization of biogeochemical cycles makes it possible to demonstrate a number of new R&D directions at the intersection of fundamental and applied researches.

Accordingly, a new practically oriented area is the biogeochemical engineering, through the development of innovative environmental technologies.

Biogeochemical technologies are the technology and technological processes based on knowledge, understanding and management of biogeochemical cycles. Scope can be related to recultivation of damaged areas, biofuel production, biogeochemical standards, risk management, etc. (Bashkin 2016).

Considering the application area of biogeochemical innovative technologies, we can highlight the following. Functioning of the gas industry in the conditions of the Far North has certain geoecological consequences in connection with journey of the equipment during the conducting exploration works, arrangement and development of new fields, gas and gas condensate production and their preparation for transport.

These geoecological consequences are expressed in mechanical impact of the equipment on a soil-vegetable cover of the territories within productive activity of the gas industry. At the same time tundra soils in parts or in full lose a vegetable cover and an organogenic layer, and the mineral horizons come to a day surface. Nevertheless, the soil has self-restoration, in the form of regeneration on its a vegetable cover over time. However in the conditions of tundra severe climate, about full self-restoration of mechanically disturbed soils it will be possible surely to judge only later decades after the beginning of this natural process. Therefore becomes extremely important to have the new ecologically oriented biogeochemical technologies for expeditious carrying out the reclamation of soil-vegetable cover of the territories damaged by technogenic influence (Galiulin et al. 2016).

The main objective of this work consisted in representation of three new ecologically oriented biogeochemical technologies of the reclamation of territories of productive activity of gas industry in the conditions of the Far North connected with use of such important and available means of reclamation as local peat.

It should be noted that these technologies already find the practical application not only in case of the reclamation of mechanically disturbed tundra soils of the Taz Peninsula (68°09' N, 76°02' E; the Yamalo-Nenets Autonomous Okrug), as area of functioning of 100% of subsidiary company of Gazprom PAS - Gazprom Dobycha Yamburg LLC, but also in case of the reclamation of disturbed and polluted soils of the Bely Island in the Kara Sea (73°15' N, 70°50' E; the Yamalo-Nenets Autonomous Okrug) where polar infrastructure for many years functions (Arabsky et al. 2015; Galiulin et al. 2016; 2016a).

Technology of Definition of Local Peat Dose Depending on Soil Full Moisture Capacity

The technology of definition of local peat dose depending on a soil full moisture capacity includes four consistently carried out stages.

So, at the first stage determine a full moisture capacity of mechanically disturbed soil by the weight method described in works (Revut 1972; Kaurichev et al. 1980). For this purpose the soil sample is previously brought to an air-dry state, crushed and sifted through a sieve with a diameter of cells of 2 mm. The essence of the analysis of a soil full moisture capacity consists in use for this glass cylinder (1) since one end the closed filter paper (2) and a gauze (3) drawn by the fixing rubber strip (4) and the soil (5) filled below the upper edge of the cylinder (Figure **1**). The cylinder with the soil prepared thus establish in a chemical glass (6) on glass sticks (7) with a length less internal diameter of a glass. Then in a glass pour the distilled water (8) so that water level matched the upper edge of the soil, and leave the cylinder in water before full saturation. Initial water level support by its filling up in process of soil absorption and water evaporation.

Full saturation of the soil by water is controlled daily pulling of the cylinder with the soil from a glass and weighing up to stabilization of weight. The full moisture capacity (FMC) of soil calculate on a formula:

$$FMC = ((a-b)/b) \cdot 100\%,$$

where a – the soil mass in cylinder after saturation by water (g), b – the mass of the air-dry soil in cylinder (g).

At the second stage select average samples of peat not less than from three peat minings for identification of peat with the maximum dehydrogenase enzyme activity and its use for soil reclamation. The dehydrogenase enzyme activity of the selected peat samples, further soils and mixture of the soil with peat analyze by the method protected by patent of the Russian Federation (Arno et al. 2013).

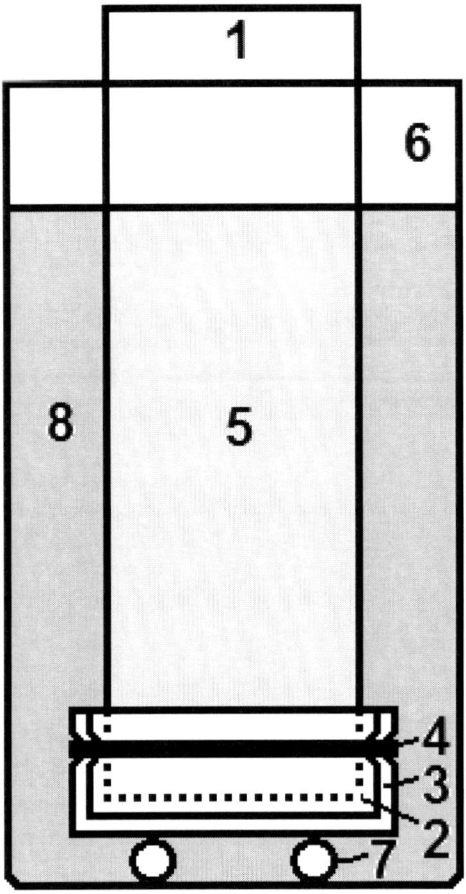

Figure 1. Device for definition of a full moisture capacity of mechanically disturbed soil: 1 – glass cylinder; 2 – paper filter; 3 - gauze; 4 – fixing rubber strip; 5 - soil; 6 – chemical glass; 7 – glass stick; 8 – distilled water.

For this purpose samples of peat place in Petri (1) cups and humidify up to 70% of a full moisture capacity and incubate in the thermostat (a) in case of 30°C within 3 days (Figure 2). Initial humidity of samples is controlled every 2 day by weighing and filling up. In 10 days select samples in 6-fold frequency for the dehydrogenase activity analysis, after bringing samples to an air-dry state directly in Petri's cups. The enzyme activity is determined by the device in the form of the modified Erlenmeyer (4) flask with a cranked shoot (5).

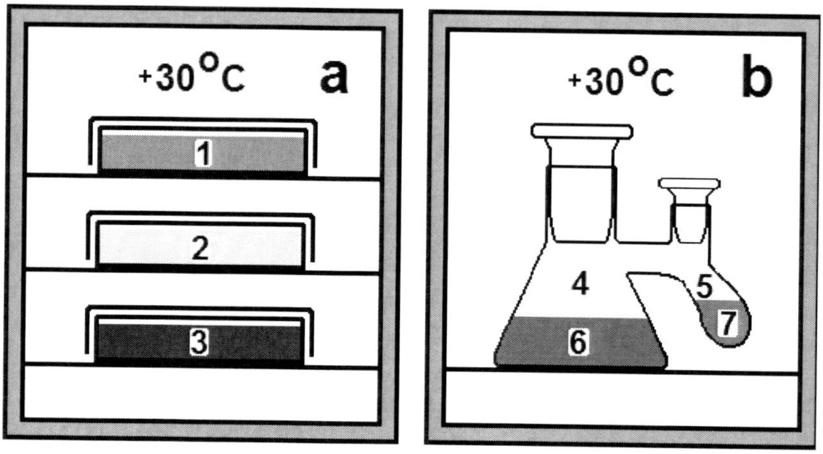

Figure 2. Equipment and device for the dehydrogenase enzyme activity analysis of various samples: a and b – thermostats; 1 – Petri's cup with peat; 2 – Petri's cup with soil; 3 - Petri's cup with soil and peat mixture; 4 – the modified Erlenmeyer's flask; 5 - cranked shoot of the flask of Erlenmeyer; 6 – reactionary mixture of a sample, calcium carbonate and water solutions of glucose and 2,3,5-triphenyltetrazoliumchloride; 7 - saturated alkaline solution of pyrogallol.

For the purpose of the anaerobic dehydrogenase activity analysis, 1 g of sample, separately from each variant, and also 0.1 g of thinly crushed calcium carbonate ($CaCO_3$), on 1 ml of 1% - water solutions of glucose ($C_6H_{12}O_6$) and 2,3,5-triphenyltetrazoliumchloride ($C_{19}H_{15}N_4Cl$) consistently place in the Erlenmeyer flask (4) and reactionary mixture (6) stir roundabouts.

By means syringe enter saturated alkaline solution of pyrogallol, $C_6H_3(OH)_3$ (7) into the cranked shoot (5) for absorption of oxygen (O_2) in the device for the purpose of creation of anaerobic conditions. Further the flask is pressurized traffic jams with use of vacuum lubricant and put in the thermostat (b) on incubation in case of 30°C for one days. It begins the biochemical reaction when 2,3,5-triphenyltetrazoliumchloride (colorless substance) accepting the hydrogen mobilized by dehydrogenase turns in the incubated environment in 2,3,5-triphenylformazan ($C_{19}H_{16}N_4$, substance of red color):

$$C_{19}H_{15}N_4Cl + H_2 = C_{19}H_{16}N_4 + HCl$$

After completion of incubation of samples extraction of a 2,3,5-triphenylformazan, which is formed in them is made from each flask by means of ethyl alcohol (C_2H_5OH) - 5 times on 4 ml. Then extracts of each sample integrate up to the volume of 25 ml and measure optical density by the spectrophotometer (wavelength λ = 490 nanometers) and calculate quantity of a 2,3,5-triphenylformazan according to the calibration graph made, for example, by from 1 to 30 mkg/ml of this substance and express in mkg of a 2,3,5-triphenylformazan/(g•day). When calculate activity of enzyme in various samples considering their humidity. The statistical proof of the maximum dehydrogenase activity of this or that sample of peat at geoecological researches of this sort will be sufficient at 5% significance value.

At the third stage choose a rational dose of the identified peat with the maximum dehydrogenase activity depending on a soil full moisture capacity in the form of a certain ratio of a peat:soil (Table 1).

At the fourth stage, carry out the soil reclamation by means peat in the conditions of *in vitro*. For this purpose soil samples without introduction of peat (2) and the soil with introduction of peat (3) within 10 days incubate at humidity of 70% of a full moisture capacity and temperature of 30°C in the thermostat (a) with the subsequent sampling in six multiple frequency (look Figure 2).

Table 1. Ratio of peat:soil depending from soil full moisture capacity

Classification of soil on level of its full moisture capacity	Full moisture capacity, %	Ratio of peat:soil
Low	40-50	1:4
	50-60	1:5
Middle	60-70	1:6
	70-80	1:7
High	80-90	1:8
	90-100	1:9

Then in these samples determine dehydrogenase activity on above described technique and with receipt of analysis results about efficiency of the soil reclamation by means of peat judge by increase in dehydrogenase activity of sample with peat introduction concerning dehydrogenase activity of sample without peat introduction.

TECHNOLOGY OF POTASSIUM HUMATE RECEIVING FROM LOCAL PEAT

The technology of receiving potassium humate as growth factor of plants from local peat includes six stages.

So, at the first stage carry out peat decalcifying by 0.1 N solution of sulfuric acid (H_2SO_4) at a ratio 1:20. The received suspension is left for 1 day and after settling the solution from a solid phase is separated a decantation, i.e., by draining of solution from sediment.

At the second stage carry out 4-5 multiple extraction (duration 20 hours) of humic acids from received sediment by 0.1 N solution of sodium hydroxide (NaOH) at ratio 1:15. Then the solid phase from alkaline solution is separated centrifugation.

At the third stage from the received alkaline solution precipitate (within 1 days) humic acids by 10% solution of hydrochloric acid (HCl) at a ratio 50:1 with the subsequent separation of sediment by centrifugation.

At the fourth stage carry out clarification of the received sediment of humic acids by dissolution in 0.5-1 l of 0.1 N of solution of sodium hydroxide (NaOH), and also sodium sulfate (Na_2SO_4) addition for coagulation of mineral particles and subsequent centrifugation of alkaline solution. Humic acids precipitate by addition of 0.1 N of solution of hydrochloric acid (HCl) before establishment pH 1-2. Then the sediment of humic acids is repeatedly washed out by the distilled water before establishment pH 6 and dried up in the thermostat at 50°C.

At the fifth stage carry out by method of nuclear-magnetic-resonant ^{13}C-spectroscopy the analysis of relative content (to %) aliphatic (0-65 ppm), polysaccharide (65-90 ppm), aromatic (90-160 ppm) and carboxyl (160-200 ppm) carbon in structural fragments of humic acids (Kalabin et al. 2000).

At the sixth stage prepare 2.5% uterine solution of potassium humate by means of addition of the distilled water and 0,1 N solution of potassium hydroxide (KOH) of with the subsequent finishing pH required solution to 7. Then from the uterine solution prepare the water solutions of potassium humate of various concentration intended for processing of seeds before crops, root top dressing and spraying during vegetation.

Technology of Combined Use of Local Peat and Potassium Humate

The technology of combined use of local peat and potassium humate includes two stages.

So, at the first stage, in the conditions of *in vitro* in special plastic vessels with a volume of 0.25 l incubate in 3-fold frequency of 100 g samples of mechanically disturbed soil without peat introduction, the soil with peat introduction, and soil with joint introduction of peat and potassium humate water solution of necessary concentration with subsequent crops and cultivation in all variants of perennial cereal grasses mixture within 30 days at day lighting in controlled hydrothermic conditions (Zhurbitsky 1968). Among perennial cereal grasses mixture can be rump beardless (*Bromus inermis*), pyreynik Siberian (*Elymus sibiricus*), tall fescue (*Festuca pratensis*), oat-ence red (*Festuca rubra*), meadow grass (*Poa pratensis*), timothy grass (*Phleum pratense*) and perennial ryegrass (*Lolium perenne*), which allow to receive dense herbage and the dense turf on the recultivating site (Sariev and Zelensky 2013; Galyamov et al. 2015). Crops of seeds of a perennial cereal grasses mixture were carried out at the rate of 30 g/m^2.

At the second stage after cutting determine elevated biomass of perennial cereal grasses mixture in various variants of experience. About efficiency of the mechanically disturbed soil reclamation in case of combined use of peat and a potassium humate judge by increase on its elevated biomass of perennial cereal grasses mixture concerning variants with the soil using only peat and also the soil without use of peat.

CONCLUSION

Thus, the essence of described new ecologically oriented biogeochemical technologies of the mechanically disturbed tundra soils reclamation consists in their step by step performance in the conditions of *in vitro* for the purpose of expeditious obtaining necessary information for carrying out the large-scale *in situ* soil reclamation.

In general these technologies allow to determine, firstly, the quantity of a rational dose of local peat necessary for carrying out reclamation of soils with various full moisture capacities, secondly, allow receiving chemically pure and stable potassium humate from local peat and finally, thirdly, soil reclamation by peat together with a potassium humate is developing.

CONFLICT OF INTEREST

The authors confirm that this article content has no conflict of interest.

ACKNOWLEDGMENT

The authors express their deep appreciation and gratitude for their affiliation institutions, namely Institute of Basic Biological Problems of Russian Academy of Sciences and also Institute of Physicochemical and Biological Problems of Soil Science of Russian Academy of Sciences for the support during research and development.

References

Arabsky, A. K., Bashkin, V. N., and Galiulin, R. V. (2015). Recultivation of disturbed tundra soils of Taz Peninsula around production activity of Gazprom Dobycha Yamburg LLC. *Ecological Annual*, **8**, 13-6.

Arno, O. B., Arabsky, A. K., Bashkin, V. N., Galiulin, R. V., Galiulina, R. A., Makliuk, O. V., and Priputina, I. V. (2013). *Control Method of Recultivation Efficiency of Disturbed Tundra Soils of Various Granulometric Composition by Dehydrogenase Activity Analysis*. RU Patent 2491137.

Bashkin V. (2016) Biogeochemical technologies for managing pollution in polar ecosystems. Environmental Pollution, Vol. 26, Switzerland, Springer.

Galiulin, R. V., Galiulina, R. A., Bashkin, V. N., and Arabsky, A. K. (2016). Disturbed tundra soils: granulometry and recultivation. *Neftegaz.RU*, **5-6**, 102-5.

Galiulin, R. V., Bashkin, V. N., Alekssev, A. O., Galiulina, R. A., and Arabsky, A. K. (2016a). Island Bely (Kara Sea): recultivation of soils. *Neftegaz.RU*, **7-8**, 96-100.

Galyamov, A. A., Gaevaya, E. V., and Zakharova, E. V. (2015). Biological recultivation of agricultural lands (reindeer pastures) on the Yamal Peninsula. *Bulletin of Krasnoyarsk State Agrarian University*, **10**, 17-22.

Kalabin, G. A., Kanitskaya, L. V., and Kushnarev, D. F. (2000) *Quantitative of nuclear-magnetic-resonance spectroscopy of natural organic raw materials and products of its conversion*. Khimiya Publishing House, Moscow, p. 408.

Kaurichev, I. S., Panov, N. P., Stratonovich, M. V., Grechin, I. P., Savich, V. I., Ganzhara, N. F., and Mershin, A. P. (1980). *Practical Work on Soil Science*. Kolos Publishing House, Moscow, p. 272.

Revut I. B. (1972). *Physics of Soils.* Kolos Publishing House, Leningrad, p.368.

Sariev, A. K. H., and Zelensky, V. M. (2013). Studying of perennial cereals grasses for biological recultivation of disturbed soils on the Yenisei North. *Achievements of Science and Engineering of AIC*, 11, pp. 27-30.

Zhurbitsky, Z. I. (1968). *Theory and Practice of Vegetative Method.* Nauka Publishing House, Moscow, p. 267.

In: Ecological and Biogeochemical Cycling … ISBN: 978-1-53612-081-3
Editor: Vladimir N. Bashkin

Chapter 9

BIOGEOCHEMICAL TECHNOLOGY FOR RECULTIVATING DISTURBED SOILS OF THE ISLAND BELY (KARA SEA)

Rauf V. Galiulin[1,*], Vladimir N. Bashkin[2,3], Andrey O. Alekseev[2], Rosa A. Galiulina[1] and Anatoly K. Arabsky[4]

[1]Institute of Basic Biological Problems of Russian Academy of Sciences, Pushchino, Russian Federation
[2]Institute of Physicochemical and Biological Problems of Soil Science of Russian Academy of Sciences, Pushchino, Russian Federation
[3]Gazprom VNIIGAZ LLC, Razvilka, Russian Federation
[4]Gazprom Dobycha Yamburg LLC, Novy Urengoy, Russian Federation

* Address correspondence to this author at the Institute of Basic Biological Problems of Russian Academy of Sciences: Pushchino, Moscow Region, 142290, Russian Federation; Tel: +7 4967 33-14-53, Fax: +7 4967 33-05-32, E-mail: rauf-galiulin@rambler.ru.

ABSTRACT

Recultivation of physically disturbed and coal polluted soils of the Island Bely (Kara Sea) is connected with using peat as the bioorganic fertilizer promoting restoration of the biogeochemical cycles of substances at the level of microbe populations which functioning is diagnosed through their biochemical activity. Efficiency of the soil recultivation, which is carried out taking into account their full moisture capacity in the conditions of *in vitro* experiment, is estimated by means of the dehydrogenase enzyme activity analysis. It presents the required information for large-scale recultivation of soils by using local peat in mixture with the disturbed soil.

Keywords: physically disturbed and coal polluted soils, full moisture capacity, recultivation, local peat, biochemical activity.

INTRODUCTION

The Island Bely (The Yamalo-Nenets Autonomous Okrug) is in the Kara Sea and separated from the Yamal Peninsula by the Strait of Malygin (Figure 1).

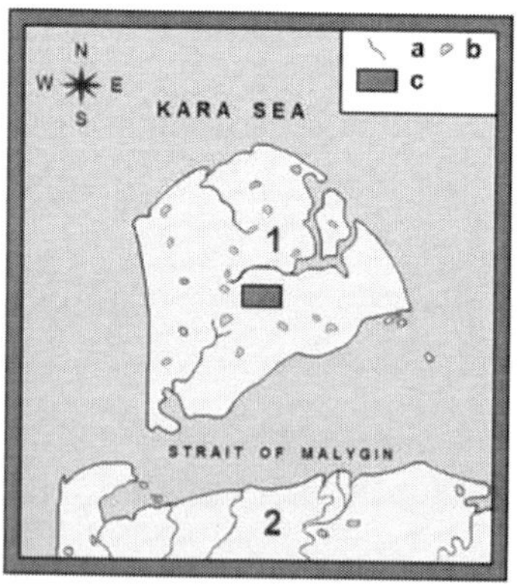

Figure 1. Map-scheme of research territory. 1 – The Island Bely (73°15' N, 70°50' E); 2 – The Yamal Peninsula; a – rivers; b – lakes; c – area on assessment of recultivating potential of physically disturbed and coal polluted soils. (https:ru.wikipedia.org/wiki/Bely_(island_Kara_sea), 2016).

The area of this island makes 1900 km^2 with hills up to 12 m, and its surface is covered with tundra vegetation. On the island there are a lot of thermo-karst lakes, which are originating in the ground depressions formed in the melting prpcesses of underground ice. The surface of this island on a relief flat, gradually going down to a south, and northern and east coasts are low and sandy, on the west and southern bank are meet breaks up to 6 m high in places. Since 1933 on the island there is the polar meteorological station named after M.V. Popov. Public attention as it is chosen to start implementation of the program for cleaning of the Arctic from garbage and waste, which has collected in the course of functioning of polar infrastructure is drawn to this island now (Tomashunas & Abakumov 2014; Vasilchuk & Vasilchuk 2015).

The examination of a part of the territory of the island conducted by us in 2015 has shown that there are sites as with physically disturbed soils that is deprived of a vegetable cover and the top organogenic layer and with the polluted soils, for example, owing to warehousing of fuel on them in the form of coal. The geoenvironmental problem demanding operational permission to avoid full degradation of a soil cover has resulted, and, therefore, to keep native flora and fauna of the polar island.

Meanwhile, the soil has the self-restoration capacity, which is happening by its gradual settling by plants, receipts of vegetable debris, development of soddy process and accumulation of a humus (Galiulin et al., 2013). However in the conditions of severe climate of the Far North a soil self-restoration will demand the long time measured decades. In this regard it is represented quite rational to accelerate restoration of a soil cover of the Island Bely by the acceptable technology of recultivation as physical disturbance and pollution of soils is among essential factors of undermining the normal biogeochemical cycles of substances in the nature. The acceptability of recultivation technology can be estimated by means of the key biochemical indicators reflecting restoration of soil fertility.

The purpose of this work was development *in vitro* conditions the recultivation technology for physically disturbed and coal polluted soils of the Island Bely, by means of use of local peat in mixture with the recultivated soil taking into account a full moisture capacity of the last.

Here the full moisture capacity is understood as that greatest quantity of moisture, which contains in the soil at full saturation of all its pores. Such methodological approach allows to obtaining necessary information for implementation in the long term of *in situ* large-scale soil recultivation by using local peat. At the same time the efficiency evaluation of soil recultivation is made by means of dehydrogenase enzyme activity analysis.

Concept of Recultivation Technology of the Island Bely Soils

The essence of the concept of recultivation technology of physically disturbed and coal polluted soils of the Island Bely consists in formation of idea about influence of a soil full moisture capacity and use of local peat for this process. It is known that the soil at full saturation by moisture of all its pores turns into the two-phase system consisting of a solid and liquid phases, except for insignificant on volume the sorbed or clamped amount of the air which has remained in the soil (Revut 1972). Use of such criterion of recultivation as full moisture capacity isn't casual as the phenomenon of a so-called hydromorphism, i.e., the temporary or constant remoistening covering a season-smelting layer when the amount of moisture exceeds 70-80% of a full moisture capacity is characteristic of the Far North soils.

As for use of peat, this organogenic rock consists not only of not completely decayed of plant remains, but also of a product of their degradation in the form of the humus including, in particular, the humic acids which are characterized by the high content of carbon and all nutritious elements, necessary for plants. From such three nutrient elements as nitrogen, phosphorus and potassium, the content of nitrogen in peat is maximal (to 3.5%). Besides peat possesses a certain pool of various physiological groups of the microorganisms (ammonifiers-aerobes, sporous bacteria, oligonitrofila, fungi, nitrifiers, denitrifiers, butyric-acid bacteria) participating in degradation of peat organic substance that makes available nutrient elements for plants (Emelyanova & Kramarenko 2004). In this regard peat introduction as bioorganic fertilizer to physically disturbed and coal polluted soils will promote restoration of a biogeochemical cycles of substances at the level of microbe populations which functioning is diagnosed through biochemical activity (Kovda 1976; Ivlev 1986).

Value of microbe populations consists not only in quantity of the delivered biomass, in one year reaches 20-50 t/ha, that approaches with land biomass of plants, and mainly in that work which they perform on a mineralization of plant remains in the soil. Dying off, microorganisms release various elements, which enter new cycles of circulation. It is remarkable also that dark color of peat promotes absorption of heat and rapid warming up of the soil that is especially important in case of its recultivation in the conditions of the Far North.

In our case, an efficiency evaluation of peat use for soil recultivation is performed by means of dehydrogenase enzyme activity analysis. This enzyme catalyzes dehydrogenation reactions, i.e., a hydrogen splitting off of the organic substances (carbohydrates, alcohols and organic acids) arriving with a plant remains to the soil and in practice is successfully used as a key biochemical indicator of restoration process of soil fertility.

Adequacy of use of dehydrogenase activity in case of an efficiency evaluation of recultivation of physically disturbed and coal polluted soils with various full moisture capacity by means of peat introduction was proved earlier by carrying out the correlation and regression analysis of experiment data on enzyme activity and full moisture capacity of tundra soils of the Taz Peninsula (The Yamalo-Nenets Autonomous Okrug) (Galiulin et al., 2013). So, calculation of the coefficient of correlation indicating the direction and degree of an associativity in variability of signs has shown existence of strong essential correlation dependence between dehydrogenase activity and a full moisture capacity of soils ($r = 0.95$). The corresponding formula of correlation dependence, i.e., the equation of linear regression allowing judging how the productive sign (y) quantitatively changes at changing factorial sign (x) on a unit of measure had the following appearance:

$$y = 7.71 + 0.15x$$

As it has appeared, the more there was a full moisture capacity of soils, the higher there was their dehydrogenase activity. The leading value of humidity for dehydrogenase activity of the soil is connected with the fact that moisture defines a normal physiological state of microorganisms and plants producing enzymes in the soil, and also supports enzymes and their substrata (carbohydrates, alcohols and organic acids) in a reactionary state.

So, with increase in a full moisture capacity of physically disturbed and coal polluted soils and introduction of peat dehydrogenase activity shall increase as such that allows using quite reasonably this indicator for an efficiency evaluation of soil recultivation.

Technology of Soil Recultivation of the Island Bely

1. Preparatory Phase for Soil Recultivation

In the territory of the Island Bely 5 samples of physically disturbed and coal polluted soils have been selected, Figure 2, (Table 1).

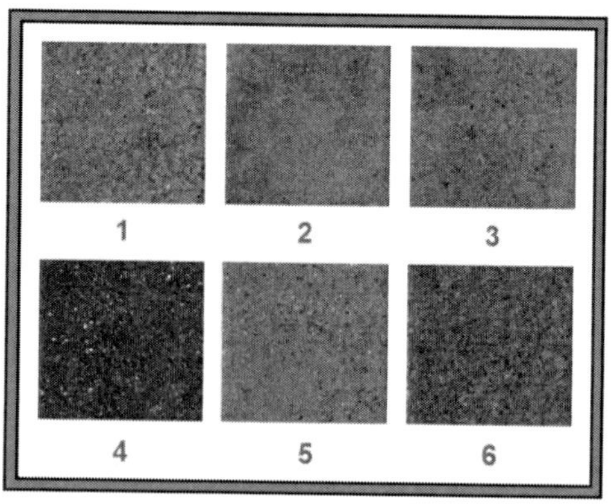

Figure 2. Color of the physically disturbed soils (1, 2, 3 and 5), coal polluted soil (4) and peat (6) from the Island Bely.

Table 1. Physicochemical properties of physically disturbed and coal polluted soils from the Island Bely

No of sample	Classification		Full moisture capacity, %	pH_{H2O}	pH_{KCl}
1	by granulometric composition	loamy sand	67.9	7.9	7.6
2		loamy sand	48.4	6.5	5.5
3		loamy sand	40.3	7.0	5.9
4		consolidated sand	49.5	6.7	6.2
5		loamy sand	44.6	8.0	7.6
6	by organogenesis	Peat	177.0	4.3	3.6

It should be noted that color of these soils is the most available, and first of all, the evident morphological feature characterizing their physical disturbance and also pollution.

From these, 4 samples on granulometric composition represented loamy sand and 1 sample – consolidated sand, which is selected from the polluted site where earlier coal was stored. Here loamy sand is understood as the content of physical clay of it, i.e., particles <0.01 mm in quantity of 10-20%, and under consolidated sand – the content of physical clay of it in quantity of 5-10%.

Soils differed also on other indicators: on the full moisture capacity, actual acidity (pH_{H2O}) and potential acidity (pH_{KCl}). Here actual acidity is understood as the acidity of soil solution created by carbonic acid (H_2CO_3), water-soluble organic acids and hydrolytic sour salts and which exerts direct impact on development of soil microorganisms and plants. Potential acidity, i.e., acidity of a soil solid phase, is meant as the acidity caused by existence of ions of hydrogen (H^+) and ions of aluminum (Al^{3+}) in the absorbed state.

Meanwhile it should be noted close connection of a soil moisture capacity with granulometric composition of the studied soils. So, easy soils, i.e., consolidated-sandy and sandy-loaming granulometric composition differ in low values of a moisture capacity, the content of a humus, plant nutrient elements and absorbing ability, and peat introduction as recultivating means, can significantly raise these indicators.

In our case we consider the use for soil recultivation of local peat, i.e., from the nearest regional areas, the Yamalo-Nenets Autonomous Okrug. It isn't casual because local peat has zonal signs proved by the analysis of content of aliphatic, aromatic, polysaccharide and carboxyl carbon atoms in structural fragments of humic acids of peat humus by method of nuclear-magnetic-resonance ^{13}C-spectroscopy (Kalabin et al., 2000). So, statistically significant distinctions in the content of some types of carbon in humic acids of a peat humus from the Yamalo-Nenets Autonomous Okrug and adjacent the Khanty-Mansi Autonomous Okrug have been established (Chumak & Sartakov 2014), Table 2.

Table 2. Comparative assessment of content of different types of carbon (%) in structural fragments of humic acids of peat humus from the Yamalo-Nenets Autonomous Okrug and the Khanty-Mansi Autonomous Okrug

Type of carbon	The Yamalo-Nenets Autonomous Okrug	The Khanty-Mansi Autonomous Okrug
Aliphatic	37.9-54.0	42.2-47.1
Aromatic	14.1-23.2	36.3-42.1
Polysaccharide	23.1-26.8	5.5-13.3
Carboxyl	7.9-10.1	4.7-8.8

These distinctions are noted, first of all, in the prevailing content of aliphatic carbon of relatively aromatic carbon in peat from the Yamalo-Nenets Autonomous Okrug as the most significant indicator of structure of a carbon skeleton of humic acids that has allowed presenting the received result as the first zone sign. The established ratio of carbon two types demonstrates the disturbed hydrophilic and hydrophobic balance as aliphatic fragments of humic acids are carriers of hydrophilic properties unlike hydrophobic aromatic fragments.

While almost identical level of aliphatic and aromatic carbon content in peat from the Khanty-Mansi Autonomous Okrug proves existence of its hydrophilic and hydrophobic balance (Sartakov 2008). Essential distinctions are noted in relatively bigger content of polysaccharide carbon in peat from the Yamalo-Nenets Autonomous Okrug, than in peat from the Khanty-Mansi Autonomous Okrug that indicates a significant role of polysaccharides in formation of humic acids in local peat that has allowed to present the received result as the second zone sign.

In general, the revealed zone signs of humic acids of peat humus from the Yamalo-Nenets Autonomous Okrug confirm its "compatibility" with a soil-vegetable cover and form a strong reason for use of local peat as bioorganic fertilizer for recultivation of physically disturbed and coal polluted soils of the Island Bely. So-called "compatibility" of local peat with a soil-vegetable cover of the Yamalo-Nenets Autonomous Okrug is caused, in particular, by universal distribution on undisturbed mineral and organogenic (peat) soils of this region of such perennial plant as the sphagnum moss (*Sphagnum*). This plant, accumulating the mineral substances arriving with rainfall and decaying upon completion of life cycle, gives them to the spreading soil together with the biomass and is an important source of formation of local peat.

2. Base Phase of Soil Recultivation

For direct carrying out recultivation of physically disturbed and coal polluted soils from the Island Bely the necessary peat:soil ratio was chosen on an indicator of a full moisture capacity from Table 3. In this case, this ratio for the soil No. 1 constituted 1:6, for other soils – 1:4. In this table as a starting settlement ratio was used the peat:soil ratio - 1:4, recommended in work (Andreev et al., 2003) for soil recultivation where friable sand from a local sand-pit, was characterized by the low level of a full moisture capacity.

Table 3. Peat:soil ratio depending from full moisture capacity of physically disturbed and coal polluted soils

Classification of soil on the level of its full moisture capacity	Full moisture capacity, %	Ratio of peat:soil
Low	40-50	1:4
	50-60	1:5
Middle	60-70	1:6
	70-80	1:7
High	80-90	1:8
	90-100	1:9

Further for confirmation of efficiency of use of mixture of local peat with the recultivated soil we carried out the comparative dehydrogenase activity analysis of this mixture with enzyme activity of physically disturbed and coal polluted soils in controlled hydrothermal conditions by the spectrophotometry method protected by the patent of the Russian Federation (Arno et al., 2013). As a result of carrying out 30 daily experiments have been established that introduction of peat stimulates dehydrogenase activity by 1.3-3.0 times depending on the soil kind that allows to speaking about a real possibility of use of peat as recultivating means (Figure 3).

CONCLUSION

Thus, realization of recultivation technology of physically disturbed and coal polluted soils from the Island Bely in the conditions of *in vitro* experiment has shown that there is an possibility of not only rational local peat using but also to proposing remediating means for the *in situ* large-scale and long term soil recultivation to accelerate a restoration of biogeochemical cycles of substances originally at the level of microbe populations, for avoiding full degradation of a soil cover as a basis of existence of native flora and fauna. Such recultivation will be an impulse to the subsequent restoration of various links of biogeochemical cycling at the levels of invertebrate organisms and vegetable biogeocenoses.

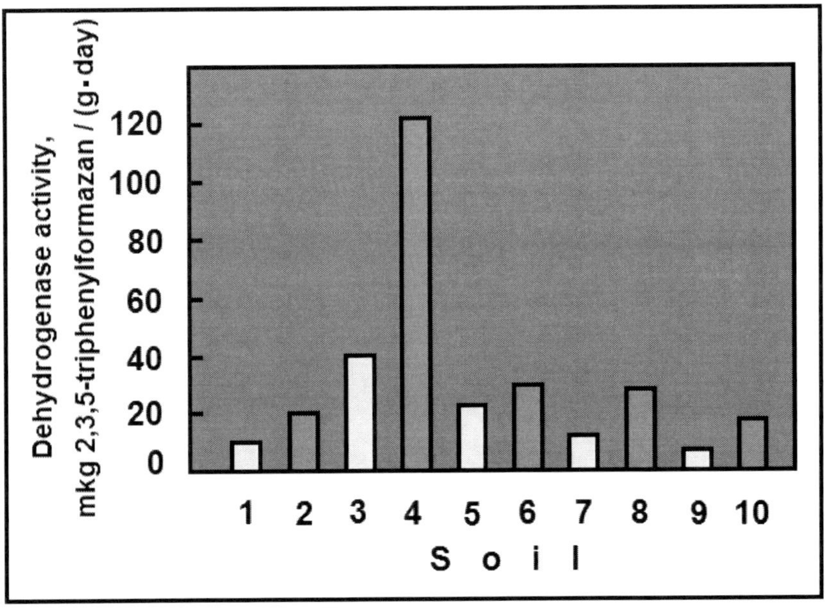

Figure 3. Dehydrogenase enzyme activity of physically disturbed and coal polluted soils from the Island Bely. 1, 3, 5, 7 and 9 – soils (control); 2, 4, 6, 8 and 10 – soils with addition of peat; 1 - peat: soil ratio is 1:6; 2, 3, 4 and 5 – 1:4.

CONFLICT OF INTEREST

The authors confirm that this article content has no conflict of interest.

ACKNOWLEDGMENTS

The authors express their deep appreciation and gratitude for their affiliation institutions, namely Institute of Basic Biological Problems of Russian Academy of Sciences and also Institute of Physicochemical and

Biological Problems of Soil Science of Russian Academy of Sciences for the support during research and development.

REFERENCES

Andreev, O. P., Stavkin, G. P., Levinzon, I. L., Perepelkin, I. B. & Lobastova, S. A. (2003). Protection and restoration of lands and landscapes of the Far North at gas production. *Ecology and Industry of the Russia*, 6, 4-9.

Arno, O. B., Arabsky, A. K., Bashkin, V. N., Galiulin, R. V., Galiulina, R. A., Makliuk, O. V. & Priputina, I. V. (2013). *Control Method of Recultivation Efficiency of Disturbed Tundra Soils of Various Granulometric Composition by Dehydrogenase Activity Analysis*. RU patent 2491137.

Chumak, V. A. & Sartakov M. P. (2014). *Experience of studying of receiving big crops conditions on the Far North (the Ostyako-Vagulsky Supporting Point, the Khanty-Mansi Experimental Station)*. AE "Technology Park of High Technologies" Publishing House, Khanty-Mansiysk, p. 514.

Emelyanova, T. Ya. & Kramarenko, V. V. (2004). Substantiation of a studying technique of peat deformation properties taking into account change of its decomposition extent. *News of the Tomsk Polytechnical University*. 5(307), p. 54-7.

Galiulin, R. V., Bashkin, V. N., Galiulina, R. A., Priputina, I. V. & Arabsky, A. K. (2013). Recultivation of the disturbed tundra soils of the Taz Peninsula by means of peat: an efficiency evaluation by means of the enzyme activity analysis. *Agrochemistry*, 4, 76-80.

Ivlev, A. M. (1986). *Biogeochemistry*. Vyshaya Shkola Publishing House, Moscow, p. 127.

Kalabin, G. A., Kanitskaya, L. V. & Kushnarev, D. F. (2000) *Quantitative of nuclear-magnetic-resonance spectroscopy of natural organic raw materials and products of its conversion*. Khimiya Publishing House, Moscow, p. 408.

Kovda, V. A. (1976). Biogeochemical cycles in the nature and their disturbance by the person. *Biogeochemical Cycles in Biosphere. Materials of the VII Plenum of SCOPE*. Nauka Publishing House, Moscow, pp. 19-85.

Revut, I. B. (1972). *Physics of Soils*. Kolos Publishing House, Leningrad, p. 368.

Sartakov, M. P. (2008). Spectroscopy of nuclear-magnetic-resonance spectroscopy ^{13}C of humic acids of peat of the Middle at Ob. *Chemistry of Vegetable Raw Materials*, 3, 135-39.

Tomashunas, V. M. & Abakumov, E. V. (2014). Content of heavy metals in soils of the Yamal Peninsula and the Island Bely. *Hygiene and Sanitation*, 6, 26-31.

Vasilchuk, A. K. & Vasilchuk, Iu. K. (2015). Engineering-geological and geochemical conditions of polygonal landscapes of the Island Bely (The Kara Sea), *Engineering Geology*, 1, 50-65.

In: Ecological and Biogeochemical Cycling … ISBN: 978-1-53612-081-3
Editor: Vladimir N. Bashkin

Chapter 10

INNOVATIVE BIOGEOCHEMICAL TECHNOLOGY FOR RECULTIVATING DISTURBED SOILS OF THE TAZ PENINSULA

***Rauf V. Galiulin*[1], *Vladimir N. Bashkin*[2,3], *Andrey O. Alekseev*[2], *Rosa A. Galiulina*[1] and *Anatoly K. Arabsky*[4]**

[1]Institute of Basic Biological Problems of Russian Academy of Sciences, Pushchino, Russian Federation
[2]Institute of Physicochemical and Biological Problems of Soil Science of Russian Academy of Sciences, Pushchino, Russian Federation
[3]Gazprom VNIIGAZ LLC, Razvilka, Russian Federation
[4]Gazprom Dobycha Yamburg LLC, Novy Urengoy, Russian Federation

ABSTRACT

The scientific substantiated innovative biogeochemical technology for recultivating disturbed soils of the Taz Peninsula (The Yamalo-Nenets Autonomous Okrug) in the impacted areas of «Gazprom Dobycha

Yamburg» LLC is presented. High efficiency of this recultivation technology is confirmed by results of the biochemical analysis of various soils and pot experiment.

Keywords: disturbed soils, innovative biogeochemical technology, recultivation, peat, perennial cereal grasses, potassium humate, biochemical analysis, dehydrogenase enzyme activity, pot experiment, grass mixture biomass

INTRODUCTION

The Taz Peninsula is in north of the West Siberian Plain (The Yamalo-Nenets Autonomous Okrug) between the Ob Bay in west and the Taz Bay in east, which both have an output to the Kara Sea (Figure 1). The surface of the peninsula is flat, covered with swamps, poorly inclined in east to the Taz Bay and falls large breaks in west to the Ob Bay. Vegetation of the peninsula is characteristic of the moss-lichen and shrubby tundra, which within centuries is used by local population for pasturable reindeer breeding. Practically all area of the tundra represents the corvine pastures including the big range of vegetable associations – moss-lichen, meadow, marsh and shrubby. At the same time the Taz Peninsula locates the unique natural gas fields on which the 100 percent of subsidiary company of "Gazprom" PAC - "Gazprom Dobycha Yamburg" LLC carries out exploration works, arrangement and development of new fields, carries out gas and gas condensate production and their preparation for transport. In the course of productive activity of this sort physical impact on a soil-vegetable cover in case of journey of the equipment connected with prospecting, well drilling and arrangement of crafts isn't excluded. As a result integrity of tundra soils is disturbed as they partially or completely lose a vegetable cover and an organogenic layer, and the mineral underlying horizons come to a day surface.

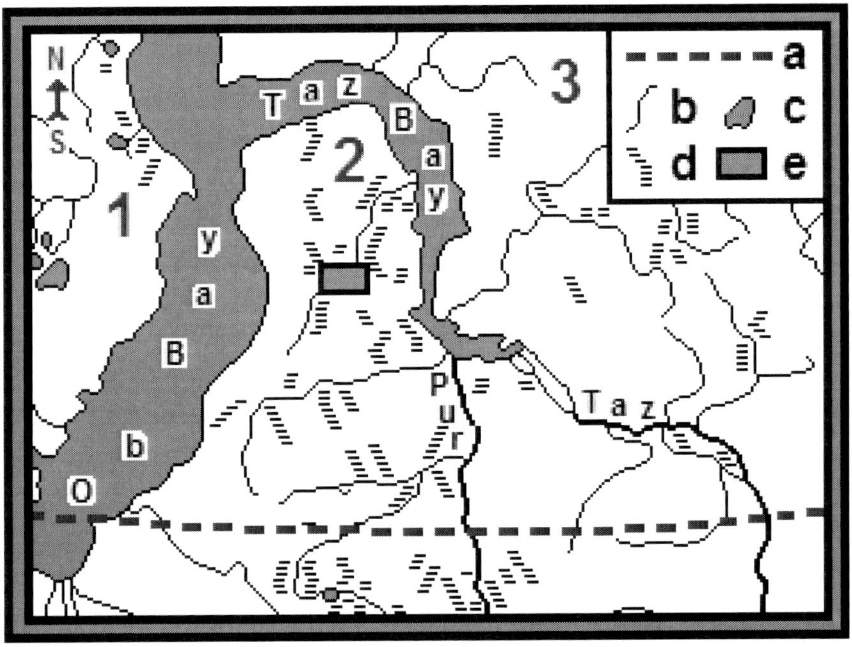

Figure 1. Map-scheme of studied territory. (1) The Yamal Peninsula; (2) The Taz Peninsula (68°09' N, 76°02' E); (3) The Gydan Peninsula; (a) The North Polar Circle, (b) rivers, (c) lakes, (d) swamps, (e) region of recultivating disturbed soils.

So, air-route observations made in the northern part of the Yenisei River have also shown the mechanical disturbance of tundra soils. For example, after laying of pipes and driving of the equipment through the route of gas pipelines continuous strips were noted: with completely destroyed mat by width of 25-50 m, with partial destruction of a moss mat and damage of shrubby vegetation by 30-70 m and with weak damage of a moss mat and preservation of shrubby vegetation by the distance of 40-100 m (Sariev & Zelensky 2013). As a result of destruction of a mat on slope sites development of erosive processes was everywhere traced, and in certain places sagging of support of the gas pipeline is revealed.

By results of assessment of a soil cover state which is carried out in the Yamal-Nenets, the Nenets and the Khanty-Mansi Autonomous Okrugs it has been revealed that the arranged transitions for deer on routes of pipelines are insufficient in a quantitative relation and often don't consider location of long-term routes of corvine herds' movement (Kapelkina & Popov 2012). Owing to insufficiency of transitions by quantity and width, the breaken up and deprived of vegetation sites are formed in places of transitions on local sites. Restoration of a vegetable cover on these sites is complicated owing to annual trampling by deer of the restoring vegetation.

For the cardinal solution of this geoenvironmental problem, on a basis of long-term researches the innovative technology for recultivating disturbed soils was developed. At present, this technology is implemented on the Taz Peninsula in the impacted areas of "Gazprom Dobycha Yamburg" LLC.

The purpose of this work consisted in presentation of the scheme of innovative biogeochemical technology for recultivating disturbed soils and confirmation of its implementation on the Taz Peninsula around functioning of "Gazprom Dobycha Yamburg" LLC.

Innovative Biogeochemical Technology for Recultivating Disturbed Soils of the Taz Peninsula

The innovative biogeochemical technology for recultivating disturbed soils consists from 3 stages including the list of consecutively carried out operations.

So, at the first stage:

- Make a large-scale map-scheme of the area of functioning of "Gazprom Dobycha Yamburg" LLC (scale 1:200000 more largely) with selection on its separate sites with the disturbed soils and measurement of their areas, and also with selection of the locations of peat deposits (Salishchev 1976);
- From the indicated sites and deposits select, respectively, average representative samples of the soil and peat from a layer of 0-5-6 cm for definition of their water-physical properties (humidity,

density and full moisture capacity) (Kaurichev et al. 1980), for the purpose of the subsequent identification of a ratio of peat:soil, necessary for recultivation of the concrete site (Table 1).

Table 1. Ratio of peat: soil for recultivation of the disturbed soil depending on the level of its full moisture capacity

Classification of soil on the level of its full moisture capacity	Full moisture capacity, %	Ratio of peat:soil
Low	40-50	1:4
	50-60	1:5
Middle	60-70	1:6
	70-80	1:7
High	80-90	1:8
	90-100	1:9

At the second stage:

- As following from the identified ratio of a peat:soil count the mass of peat sealing in 5-6 cm layer of the disturbed soil, and also mass of the itself disturbed soil in a layer of 5-6 cm for the area of the recultivating site;
- The mass of peat is previously brought to a friable state by air drying that is necessary for convenience of its uniform distribution on all site area and further sealing in a layer of the disturbed soil;
- Seal of peat in 5-6 cm a layer of the disturbed soil and crops of seeds of mixture of perennial cereal grasses carry out by the principle of the device of lawns on big squares or the "regrassing" method, using appropriate technologies and the equipment (Stefanovich 2010; Galyamov et al. 2015);
- Among perennial cereal grasses can be rump beardless (*Bromus inermis*), pyreynik Siberian (*Elymus sibiricus*), tall fescue (*Festuca pratensis*), oat-ence red (*Festuca rubra*), meadow grass (*Poa pratensis*), timothy grass (*Phleum pratense*) and perennial ryegrass (*Lolium perenne*), which allow to receive dense herbage and the dense turf on the recultivating site; effective method of increase in

stability the creating of phytocenosis at recultivating disturbed soils is crops of indicated grass mixture with inclusion of local plant types (Kapelkina & Popov 2012; Sariev & Zelensky 2013; Galyamov et al. 2015);

At the third stage:

- For improvement of sowing properties of seeds, regulation of plant state at various stages of development and in the course of forming of their productivity, and also increase of plant resistance to adverse effects of the external environment apply the potassium humate preparation used in certain doses for processing of seeds before crops, root top dressing and spraying during vegetation with use of appropriate technology and the equipment;
- A potassium humate preparation is received by the original method from local peat (The Yamalo-Nenets Autonomous Okrug) when extraction, first of all, of humic acids from peat and their cleaning is made in due form the producing of chemically pure substances which is almost not mentioning molecular structures of these acids that, eventually, guarantees creation of stable potassium humate preparation.

This method includes the following stages:

1) decalcifying of peat by 0.1 N solution of sulfuric acid (H_2SO_4) at a ratio 1:20; then the received suspension is left for 1 days, and after its upholding solution from a solid phase is separated by decantation, i.e., by draining of solution from a sediment;
2) carrying out 4-5 multiple extraction (lasting up to 20 hours) of humic acids from received sediment by 0.1 N solution of the sodium hydroxide (NaOH) at a ratio 1:15; then the solid phase is separated from alkaline solution by centrifugation;
3) sedimentation (within 1 days) humic acids from the received alkaline solution by 10% solution of hydrochloric acid (HCl) at a

ratio 50:1 with the subsequent separation of a sediment by centrifugation;

4) clarification of the received sediment of humic acids by dissolution in 0.5-1.0 l of 0.1 N of solution of sodium hydroxide and also addition of sodium sulfate (Na_2SO_4) for coagulation of mineral particles and the subsequent centrifugation of alkaline solution; then humic acids precipitate by addition 0.1 N of solution of hydrochloric acid before establishment pH 1-2; the received sediment of humic acids is repeatedly washed out the distilled water before establishment pH 6 and dried up in the thermostat at 50°C;
5) preparation from the received powder of humic acids of uterine solution of a potassium humate of necessary concentration, by means of addition in the corresponding sample of 0.1 N of solution of potassium hydroxide (KOH) and the distilled water with the subsequent finishing of pH required solution to the value equal to 7 units; then from uterine solution prepare the water solutions of a potassium humate of various concentration intended for processing of seeds before crops, root top dressing and spraying during vegetation;

- and finally, further care of vegetation on the recultivating site is carried out, also using appropriate technologies and the equipment; at blindness in grasses sward carry out an additional sowing of seeds of perennial cereal grasses (Stefanovich 2010).

Implementation of Innovative Biogeochemical Technology for Recultivating Disturbed Soils of the Taz Peninsula

On first stage for confirmation of the fact of implementation of innovative biogeochemical technology for recultivating disturbed soils on the Taz Peninsula the biochemical analysis in various samples which are quickly delivered from the background territory (gleyzem of cryogen-ferruginous tundra), the site planned for recultivation (the disturbed soil), the peat deposit (peat) and also the region of recultivation (the recultivating soil) was carried out (Khrenov 2011).

The essence of the biochemical analysis consisted in determination of one of key indicators of fertility of the soil – the dehydrogenase enzyme activity in dynamics (5, 10 and 15 days), using the method protected by two patents of the Russian Federation (Bashkin et al. 2010; Arno et al. 2013). It isn't casual as the biochemical analysis differs in high sensitivity and express, sufficient accuracy and stability of indicators in comparison with the existing methods of the microbiological analysis.

As for a dehydrogenase, this enzyme catalyzes dehydrogenation reaction, i.e., a hydrogen splitting off, of organic substances, in particular, carbohydrates, alcohols and organic acids coming with the vegetable remains to the soil and positively correlates with a number of microorganisms, accurately reflecting their functional activity.

It has been established that the studied soils sharply differ on the content of organic carbon as other key indicator of fertility of the soil (Kaurichev et al. 1980), Table 2.

Table 2. Dynamics of dehydrogenase activity (mkg 2,3,5-triphenylformazan/(g/day) of various soils and peat of the Taz Peninsula

Soil	Content of organic carbon, %	Day		
		5	10	15
Gleyzem of cryogen-ferruginous tundra	1.7	50.7	66.4	51.2
Disturbed soil	0.2	5.6	12.7	8.5
Peat from deposit	49.2	156.3	196.4	170.0
Recultivated soil	16.1	320.4	341.9	437.0

At the same time the maximum content of organic carbon characterized peat and the minimum content of this substance characterized the disturbed soil. The biochemical analysis has shown that the lowest dehydrogenase activity is observed at the disturbed soil, i.e., it was 5.2-9.1

times less, than at gleyzem of cryogen-ferruginous tundra, used in our researches as a standard (Khrenov 2011).

Introduction of peat into the disturbed soil and use of a potassium humate preparation at cultivation of perennial cereal grasses had to lead to increase in activity of enzyme. Accordingly, dehydrogenase activity of the recultivating soil in comparison with the disturbed soil was increased by 26.9-57.2 times. The fact that peat enzyme activity was higher than activity of gleyzem of cryogen-ferruginous tundra by 3.0-3.3 times demonstrates the stimulating effect of peat. It is symptomatic that in all terms of observation dehydrogenase activity positively correlated with the content of organic carbon that is convincing confirmation of a correctness of results of researches.

At the second stage, it has been carried out assessment of recultivation potential of the disturbed soil, by means of statement of vegetative experiment with cultivation of the grass mixture including oat-ence red (*Festuca rubra*), tall fescue (*Festuca pratensis*), meadow grass (*Poa pratensis*), perennial ryegrass (*Lolium perenne*) and timothy grass (*Pheleum pratense*) as recommended (Zhurbitsky 1968). The present grass mixture grew up on the disturbed soil (control), the disturbed soil with peat addition in the ratio a peat:soil - 1:6 defined depending on its full moisture capacity (Table 1), and also on the disturbed soil with peat addition in the ratio - 1:6 and 0.125% water solution of potassium humate. Humidity of samples in vegetative vessels was more than 70% of a full moisture capacity, i.e., characteristic for conditions so-called of a hydromorphism of a season-thawing layer of tundra soils. Crops of seeds of a grass mixture were carried out at the rate of 30 g/m^2.

Results of vegetative experiment lasting 2 weeks were shown that biomass of plants has accurately expressed tendency to increase from the disturbed soil through the disturbed soil with addition of peat and to the disturbed soil with joint use of peat and a potassium humate (Figure 2).

If the biomass of grass mixture at addition to the soil of peat increases by 15%, then with addition of peat and a potassium humate by 52% concerning control that confirm high potential of recultivation of the disturbed soil by means of peat and a potassium humate.

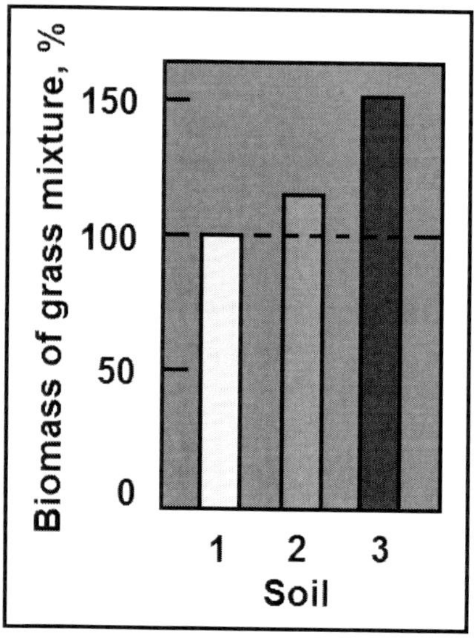

Figure 2. Biomass of grass mixture (%) on disturbed soil - control (1), disturbed soil with addition of peat in ratio peat:soil - 1:6 (2) and disturbed soil with addition of peat in ratio peat:soil - 1:6 and 0.125% water solution of potassium humate (3).

CONCLUSION

Thus, data on the biochemical analysis of the researched soils and pot experiment confirm the fact of implementation of innovative biogeochemical technology for recultivating disturbed tundra soils of the Taz Peninsula. In general these experimental data allow coming to conclusion about visibility of further broad-scale application of the given technology for reclamation of disturbed and polluted soils in the impacted territories of «Gazprom Dobycha Yamburg» LLC: areas of exploration works, arrangement and development of new fields, gas and gas condensate production and preparation for transport.

Conflict of Interest

The authors confirm that this article content has no conflict of interest.

Acknowledgments

The authors express their deep appreciation and gratitude for their affiliation institutions, namely Institute of Basic Biological Problems of Russian Academy of Sciences and also Institute of Physicochemical and Biological Problems of Soil Science of Russian Academy of Sciences for the support during research and development.

References

Arno, O. B., Arabsky, A. K., Bashkin, V. N., Galiulin, R. V., Galiulina, R. A., Makliuk, O. V. & Priputina, I.V. (2013). Control Method of Recultivation Efficiency of Disturbed Tundra Soils of Various Granulometric Composition by Dehydrogenase Activity Analysis. RU Patent 2491137.

Bashkin, V. N., Bukhgalter, E. B., Galiulin, R. V., Konyaev, S. V., Kalinina, I. E. & Galiulina, R. A. (2010). Control Method for Cleaning of Soils Contaminated by Hydrocarbons and Neutralization of Hydrocarbons Sludge by Dehydrogenase Activity Analysis. RU Patent 2387996.

Galyamov, A. A., Gaevaya, E. V. & Zakharova, E. V. (2015). Biological recultivation of agricultural lands (reindeer pastures) on the Yamal Peninsula. *Bulletin of Krasnoyarsk State Agrarian University*, 10, 17-22.

Kapelkina, L. P. & Popov, A. I. (2012). Assessment of state and recultivation of disturbed lands on oil field of the North of Russia. *Bulletin of Orenburg State Agrarian University*, 37-1 (5), 211-4.

Kaurichev, I. S., Panov, N. P., Stratonovich, M. V., Grechin, I. P., Savich, V. I., Ganzhara, N. F. & Mershin, A. P. (1980). *Practical Work on Soil Science*. Kolos Publishing House, Moscow, p. 272.

Khrenov, V. Ya. (2011). *Soils of Cryolithozone West Siberian: Morphology, Physicochemical Properties, Geochemistry,* Nauka Publishing House, Novosibirsk, p. 211.

Salishchev, K. A. (1976). *Map-science*. MGU Publishing House, Moscow, p. 438.

Sariev, A. Kh. & Zelensky, V. M. (2013). Studying of perennial cereals grasses for biological recultivation of disturbed soils on the Yenisei North. *Achievements of Science and Engineering of AIC*, 11, pp. 27-30.

Stefanovich, G. S. (2010). Device and maintenance of sowned lawns from local cereals in the Urals Region. *Agrarian Bulletin of the Urals*, 11-1 (77), 22-4.

Zhurbitsky, Z. I. (1968). *Theory and Practice of Vegetative Method*. Nauka Publishing House, Moscow, p. 267.

In: Ecological and Biogeochemical Cycling … ISBN: 978-1-53612-081-3
Editor: Vladimir N. Bashkin

Chapter 11

THE ROLE OF PLANT GROWTH-PROMOTING RHIZOBACTERIA IN THE REMEDIATION OF PB-CONTAMINATED SOIL

Valery P. Shabayev*

Institute of Physicochemical and Biological Problems in Soil Science, Russian Academy of Sciences, Pushchino, Russia

ABSTRACT

A series of pot experiments on gray forest soil artificially contaminated with water-soluble lead compounds has investigated the effect of plant growth-promoting *Pseudomonas* rhizobacteria on barley growth, uptake of lead by barley plants and its distribution in soil as a constituent of various compounds.

Inoculation with the bacteria significantly reduced the concentration of lead in shoots and roots in the first half of the growing season (up to the heading stage). A decrease in Pb uptake by inoculated plants corresponded to an increased binding of lead in soil in the form of extracted NH_4Oac complexes and organic compounds in the first half of the growing season. Thereby, the mobility of metal was reduced, without

* Corresponding author Email: VPSH@rambler.ru.

changing the reaction of the soil medium, probably due to bacterial production of organic metabolites, such as siderophores.

Application of strain *P. fluorescens* 21 eliminated Pb toxicity; herewith, the plant weight (grain weight including) increased up to the level of the uncontaminated uninoculated control. Other possible mechanisms of plant resistance to the toxic effect of lead at the introduction of the bacteria are discussed.

Keywords: *pseudomonas*, plant weight, Pb in plants, speciation of Pb in soil

INTRODUCTION

Lead, along with other heavy metals, is a hazardous ecological pollutant, especially in areas of high anthropogenic pressure (Islam et al. 2007). It enters soil mainly as a result of industrial and mining enterprises' activities, combustion of fossil fuels and production of alkaline storage batteries (Khan et al. 2009). Application of agrochemicals and long-time use of wastewater residues as fertilizers also add significant amounts of heavy metals, lead including, to soil (McGrath et al. 1995). Increased Pb concentrations inhibit many metabolic processes in various cell components and result in weaker plant growth (Sharma and Dubey 2005).

One of the ways to decrease the phytotoxicity of heavy metals and to remediate polluted soils is the use of plant growth-promoting rhizobacteria (PGPR) (Lebeau et al. 2008; Gamalero et al. 2009; Khan et al. 2009). Microorganisms associated with plant roots can have a significant effect on mineral nutrition and resistance of plants to unfavourable factors owing to numerous mechanisms involved in plant–microbe interactions. PGPR can modify plant growth and development by producing antibiotics to suppress soil pathogens, enhancing the uptake of nutrients and water by plants and, thus, increasing plant biomass, owing to the formation of growth hormones, e.g., indolylacetic acid (IAA), gibberellines and cytokinins, as well as to other mechanisms, such as the formation of siderophores, dissolution of phosphorus, fixation of atmospheric nitrogen and protection of plants from the toxicity of heavy metals (Lebeau et al. 2008).

Soil is a natural barrier on the way of heavy metals' migration to adjacent media and uptake by plants. It is the barrier functions of soils that determine a significant decrease in the entry of heavy metals, lead including, into plants. The major factors in the inactivation of heavy metals in soils are their binding in low-mobile compounds (Minkina et al. 2009) and the reaction of the soil medium (Khan et al. 2009). Microorganisms play an important role in the fate of toxic metals in the environment through the agency of numerous mechanisms affecting transformations between soluble and insoluble forms (Gadd 2000). Because of this, microorganisms can enhance the barrier functions of soil with respect to heavy metals and reduce their toxic action on plants as the result of being bound in various compounds of reduced mobility.

The soil mechanisms of plant resistance to the negative effect of heavy metals at the introduction of PGPR, in particular, the way the bacteria affect the speciation and distribution of heavy metals, lead including, in soils have been little investigated. The aims of this study were (1) to determine if introduction of *Pseudomonas* plant growth-promoting rhizobacteria, which improve mineral nutrition and yield of various agricultural crops, can contribute to the growth of barley plants on a soil artificially contaminated with water-soluble Pb compounds; (2) to assess the Pb uptake by PGPR-inoculated plants; (3) to investigate the effect of applied bacteria on Pb speciation in soil and on soil pH.

METHODS

Greenhouse Pot Experiments

A series of greenhouse pot experiments was carried out to grow barley plants (*H. vulgare* L., cv. Moskovsky 2), inoculated with *Pseudomonas* plant growth-promoting rhizobacteria, on soil artificially contaminated with water-soluble lead compounds.

Bacterial Strains and Inoculation Procedure

The following strains were used: strain *P. fluorescens* 20 increasing the yield of soybean (Shabayev et al. 1996), winter wheat (Shabayev and Smolin 2000; Shabayev and Voronina 2007), oat (Shabaev 2004), radish (Shabayev et al. 1998) and food beet (Shabayev and Smolin 2002); strain *P. fluorescens* 21 increasing the yield of soybean (Shabayev et al. 1996), winter wheat (Shabayev and Smolin 2000) and strain *P. putida* 23 increasing the yield of red beet (Mineev et al. 1991), winter wheat (Shabayev and Smolin 2000; Shabayev and Voronina 2007), oat and barley (Shabaev 2004). Sterilized seeds were placed on the surface of soil in the pot, inoculated with aqueous suspensions of the bacteria at a rate of 10^8 colony-forming units per plant and covered with a 3-cm-thick layer of soil. Adequate amounts of autoclaved bacterial suspensions were used in variants without inoculation.

Soil

Gray forest loamy soil was used. Samples were taken at an abandoned field (from a depth of 0–20 cm) in southern Moscow Region, Russia. Wet soil was thoroughly mixed and passed through a 1-cm sieve. The chemical characteristics of soil in the experiments varied: pH within the range of 5.06–6.10; organic C, 0.96–1.15%; total N, 0.11–0.13%; P and K (0.2 N HCl), 42–87 and 86–220 mg/kg soil, respectively.

Experiments

Experiment 1

In pots containing 200 g of Pb-contaminated soil each, five plants in each pot were grown up to the tillering stage for 14 days. In the first variant, strain *P. fluorescens* 20 was added; in the second, strain *P. putida* 23. There was also a variant without bacterial inoculation.

Experiment 2

In pots containing 500 g of Pb-contaminated soil each, six plants in each pot were cultivated up to the stooling phase for 30 days. There were variants with and without inoculation with *P. fluorescens* 21.

Experiment 3

In pots filled with 800 g of Pb-contaminated soil each, five plants in each pot were grown up to the heading stage for 45 days. Strain *P. fluorescens* 20 in one variant and strain *P. fluorescens* 21 in the other were introduced into to the contaminated soil. Variants with and without contamination in the absence of bacterial inoculation were also used.

Experiments 4 and 5

In both experiments, in pots containing 5 kg of Pb-contaminated soil each, 12 plants in each pot were cultivated with and without application of *P. fluorescens* 21. In Experiment 5, there was a variant without Pb contamination without application of the bacterium. In Experiment 4, plants were cut at the heading stage and at maturity; in Experiment 5, plants were grown to maturity.

Each pot was watered up to 60% WHC of soil. Treatments of variants were repeated three times in Experiments 1 and 2, and four times in Experiments 3 to 5.

Application of Lead and Mineral Fertilizers

In Experiment 1, 300 mg of lead as $Pb(CH_3COO)_2$ (Reakhim, Russia) per kg of soil; in Experiments 2–5, 200 mg of lead as $Pb(NO_3)_2$ (Merck, Germany) per kg of soil was thoroughly mixed with the soil volume in the pot 10 days before seed sowing and inoculation. In Experiments 1–3 and 5, soils were fertilized with N, P and K in terms of 80 mg each; in Experiment 4, in terms of 100 mg each element per kg of soil as NH_4NO_3, KH_2PO_4 and K_2SO_4 simultaneously with application of lead. In Experiments 3 and 5, treatments with $Pb(NO_3)_2$ were adjusted to the same level of N as in the corresponding variants of uncontaminated and N-fertilized soil.

Preparation and Analyses of Plant and Soil Samples

After cutting plants and taking soil samples, roots were separated from soil by tap water and washed with distilled water. Shoots, grain, straw (chaff including) and roots were dried at 70°C, weighed and ground to uniform consistency. Air-dried soil samples were ground and passed through a 1-mm sieve. A 1-g plant sample was digested in a 20-ml mixture of concentrated HNO_3 and $HClO_4$ (2:1 v/v) and analyzed for the content of lead. Experiments 2–4 determined the lead fraction extracted from soil with 1 N CH_3COONH_4 (pH 4.8) after shaking 3 g soil with 30 ml reagent. Pb speciation in soil in Experiment 3 was determined by sequential fractionation (Ladonin 2006). An exchangeable Pb fraction (extractant, $Ca(NO_3)_2$) and fractions associated with carbonates (CH_3COOH), organic matter ($K_4P_2O_7$) and iron minerals (Tamm's reagent) were isolated after sequentially shaking 2 g soil with 20 ml of each reagent. To determine the content of metal in the fraction bound to clay minerals, the soil residue after fractionation was decomposed by chlorazotic acid. An Optima 5900 DV inductively coupled plasma optical emission spectrometer (Perkin Elmer, USA) was used to measure the concentration of heavy metal in solutions. The pH of the soil suspension was assessed by a pH 211 pH meter (HANNA Instruments, Germany) after shaking 20 g soil with 50 ml of 1 N KCl.

Statistical Analysis

The data of all parameters for each variant were expressed as the means ± standard deviations. Significant differences between inoculated variants and uninoculated control were determined using Student's *t*-test ($\alpha = 0.05$).

RESULTS

Plant Weight and Pb Concentration

Table 1 shows the dry weight of and Pb content in barley plants in the first half of the vegetation period (up to and including the heading stage of plants). In Experiment 1, in the tillering stage the Pb concentrations in shoots of plants inoculated with *P. putida* 23 or *P. fluorescens* 20 under lead stress conditions were, respectively, almost two and three times lower than in the uninoculated control. Application of *P. fluorescens* 21 in Experiment 2 reduced the content of Pb in barley shoots and roots in the stooling stage by 28–29%. In the early periods of plant growth in the above experiments, there was no noticeable difference in the weights of plants untreated and treated with bacteria. The most significant decrease in the concentration of lead in shoots under the influence of bacterial inoculations was observed in the heading stage in Experiment 3. The use of *P. fluorescens* 20 and *P. fluorescens* 21 reduced the Pb concentration in shoots 3.5- and 7-fold, respectively, in comparison with the elevated concentration of metal in the uninoculated variant. The presence of *P. fluorescens* 21 also reduced the Pb content in roots by 23%.

Experiment 3 showed only a slight decrease in shoot weight during the heading stage in metal-contaminated soil without bacterial inoculation. At a lead stress, barley plants inoculated with *P. fluorescens* 21 had larger shoot and root weights as compared with the uninoculated variant. At the same time, the weights of shoots and roots of Pb-contaminated plants inoculated with this strain, respectively, rose as a tendency (shoots) and increased significantly (roots) as compared with the variant without contamination without inoculation. Growth response induced by inoculation with *P. fluorescens* 20 was less than for *P. fluorescens* 21. Similar dependences for *P. fluorescens* 21 were observed in the heading stage in Experiment 4. The use of the bacterium on Pb-treated soil in this experiment increased the weight of shoots 1.6-fold and of roots by 22% versus the uninoculated control. Herewith, the content of lead in shoots of inoculated plants was significantly (1.5-fold) lower than that of uninoculated plants. In roots this index decreased significantly, too, but to a lower extent than in shoots.

Table 1. Plant weight and Pb content of barley in the first half of the growing season

Experiment Number	Variant	Pb rate	Growth stage	Plant organ	Dry matter	Pb content
		mg kg^{-1}			g pot^{-1}	mg kg^{-1}
1	Pb without inoculation	300	tillering	shoots	0.14 ± 0.03	34 ± 5
	Pb + *P. fluorescens* 20				0.13 ± 0.02	11 ± 3[a]
	Pb + *P. putida* 23				0.14 ± 0.03	18 ± 4[a]
2	Pb without inoculation	200	stooling	shoots	1.51 ± 0.30	7 ± 2
	Pb + *P. fluorescens* 21				1.54 ± 0.21	5 ± 1[a]
	Pb without inoculation			roots	0.37 ± 0.06	333 ± 41
	Pb + *P. fluorescens* 21				0.35 ± 0.05	235 ± 29[a]
3	without Pb without inoculation	200	heading	shoots	2.16 ± 0.28	ND
	Pb without inoculation				1.99 ± 0.15	7 ± 2
	Pb + *P. fluorescens* 20				2.13 ± 0.17	2 ± 1[a]
	Pb + *P. fluorescens* 21				2.33 ± 0.30[a]	1 ± 1[a]
	without Pb without inoculation			roots	0.52 ± 0.09	6 ± 1[a]
	Pb without inoculation				0.54 ± 0.10	188 ± 17
	Pb + *P. fluorescens* 20				0.53 ± 0.08	173 ± 14
	Pb + *P. fluorescens* 21				0.64 ± 0.08[a]	144 ± 15[a]
4	Pb without inoculation	200	heading	shoots	2.20 ± 0.49	16 ± 3
	Pb + *P. fluorescens* 21				3.59 ± 0.69[a]	11 ± 2[a]
	Pb without inoculation			roots	0.23 ± 0.04	724 ± 120
	Pb + *P. fluorescens* 21				0.28 ± 0.05[a]	598 ± 101[a]

Values are means ± standard deviations; ND, non-detectable.

[a] Significantly different from the corresponding variant with Pb without inoculation.

Table 2. Plant weight and Pb content of barley at maturity

Experiment Number	Variant	Pb rate mg kg^{-1}	Plant organ	Dry matter g pot^{-1}	Pb content mg kg^{-1}
4	Pb without inoculation	200	grain	3.74 ± 0.63	1 ± 1
	Pb + *P. fluorescens* 21			5.52 ± 0.92[a]	1 ± 1
	Pb without inoculation		straw	11.86 ± 2.01	11 ± 2
	Pb + *P. fluorescens* 21			12.79 ± 3.09	10 ± 2
	Pb without inoculation		roots	0.73 ± 0.25	460 ± 69
	Pb + *P. fluorescens* 21			0.82 ± 0.30	415 ± 79
5	without Pb without inoculation	200	grain	10.75 ± 1.75[a]	ND
	Pb without inoculation			7.40 ± 0.99	1 ± 1
	Pb + *P. fluorescens* 21			10.55 ± 1.69[a]	1 ± 0
	without Pb without inoculation		straw	13.12 ± 2.51	ND
	Pb without inoculation			10.65 ± 1.69	3 ± 1
	Pb + *P. fluorescens* 21			12.78 ± 2.07	2 ± 1
	without Pb without inoculation		roots	2.20 ± 0.50[a]	6 ± 3[a]
	Pb without inoculation			1.54 ± 0.49	335 ± 0.61
	Pb + *P. fluorescens* 21			1.85 ± 0.63[a]	241 ± 0.45[a]

Values are means ± standard deviations, ND, non-detectable.

[a] Significantly different from the corresponding variant with Pb without inoculation

Mature Plant Weight and Pb Concentration

The positive effect of *P. fluorescens* 21 on barley growth on Pb-treated soil was retained up to maturity of plants in Experiments 4 and 5 (Table **2**).

Herewith, the toxicity of metal had an effect only on uninoculated plants (Experiment 5). Application of the bacterium under lead stress conditions increased grain weight 1.4–1.5-fold in both experiments as compared with the uninoculated variants, whereas the weight of straw showed only a tendency to an increase in both experiments, and the weight of roots increased significantly only in Experiment 5. Besides, against the background of a decrease in the weight of uninoculated plants on Pb-contaminated soil in Experiment 5, application of strain 21 under contaminated conditions contributed to the production of the same weight of barley, grain including, as in the variant without contamination and without inoculation. Pb concentrations decreased insignificantly in above-ground organs of mature barley, grain including, after inoculation with strain 21. Herewith, grain and straw of inoculated plants were found to have insignificant concentrations of lead. Roots contained an average of 2 orders of magnitude more Pb than grain and straw. The content of Pb in roots in the presence of *P. fluorescens* 21 tended to decrease in Experiment 4 and reduced significantly in Experiment 5.

Table 3. Pb content in a soil fraction extracted with ammonium acetate after cutting barley plants

Experiment Number	Variant	Growth stage	Pb content in soil mg kg^{-1}
2	Pb without inoculation	Stooling	47 ± 9
	Pb + *P. fluorescens* 20		70 ± 13[a]
3	Pb without inoculation	Heading	24 ± 5
	Pb + *P. fluorescens* 20		39 ± 5[a]
	Pb + *P. fluorescens* 21		35 ± 6[a]

Values are means ± standard deviations.

[a] Significantly different from the corresponding variant with Pb without inoculation

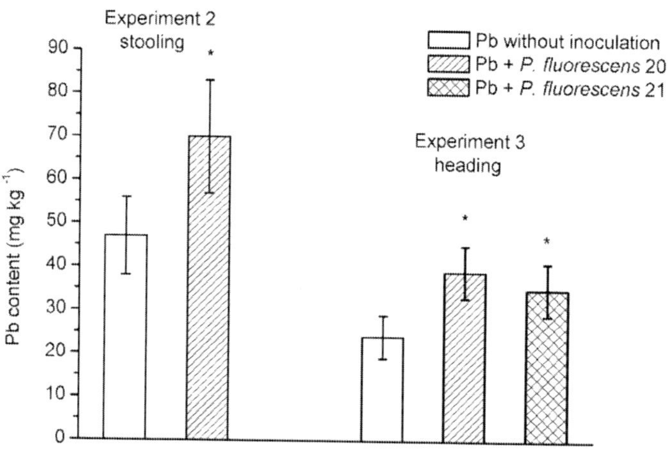

Figure 1. Pb content in a soil fraction extracted with NH_4OAc after cutting barley plants. Values are means ± standard deviations ($n = 4$). Asterisked data denote the values that differ significantly from those for the treatment with Pb without inoculation ($P < 0.05$).

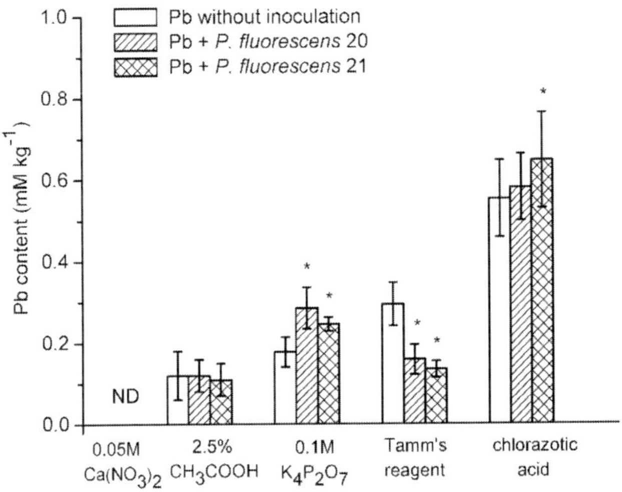

Figure 2. Pb fractions successively extracted from soil with different reagents at the heading stage of barley plants in Experiment 3. Values are means ± standard deviations (n = 4). ND, non-detectable. Asterisked data denote the values that differ significantly from those for the treatment with Pb without inoculation ($P < 0.05$).

Table 4. Pb fractions extracted from soil by the method of sequential extracts. Experiment 3

Growth stage	Variant	Extractant	Pb fraction	Pb mM kg^{-1}
Heading	Pb without inoculation	0.05 M $Ca(NO_3)_2$	exchangeable	ND
	Pb + *P. fluorescens* 20			ND
	Pb + *P. fluorescens* 21			ND
	Pb without inoculation	2.5% CH_3COOH	specifically absorbed	0.012 ± 0.006
	Pb + *P. fluorescens* 20			0.012 ± 0.004
	Pb + *P. fluorescens* 21			0.011 ± 0.004
	Pb without inoculation	0.1 M $K_4P_2O_7$	bound to organic matter	0.178 ± 0.037
	Pb + *P. fluorescens* 20			0.285 ± 0.051[a]
	Pb + *P. fluorescens* 21			0.246 ± 0.017[a]
	Pb without inoculation	Tamm's reagent	bound to amorphous Fe oxides and hydroxides	0.294 ± 0.053
	Pb + *P. fluorescens* 20			0.159 ± 0.037[a]
	Pb + *P. fluorescens* 21			0.135 ± 0.020[a]
	Pb without inoculation	residual	bound to clay minerals	0.553 ± 0.094
	Pb + *P. fluorescens* 20			0.581 ± 0.081
	Pb + *P. fluorescens* 21			0.647 ± 0.117[a]

Values are means ± standard deviations; ND, non-detectable.

[a] Significantly different from the corresponding variant with Pb without inoculation

Table 5. Soil pH after cutting barley plants

Experiment number	Variant	Growth stage	pH
2	Pb without inoculation	stooling	6.55 ± 0.16
	Pb + *P. fluorescens* 20		6.54 ± 0.11
3	without Pb without inoculation	heading	5.18 ± 0.10
	Pb without inoculation		5.36 ± 0.11
	Pb + *P. fluorescens* 20		5.42 ± 0.15
	Pb + *P. fluorescens* 21		5.40 ± 0.13
4	Pb without inoculation	heading	5.72 ± 0.11
	Pb + *P. fluorescens* 21		5.72 ± 0.11
	Pb without inoculation	maturity	6.02 ± 0.13
	Pb + *P. fluorescens* 21		6.22 ± 0.16
5	without Pb without inoculation	maturity	6.57 ± 0.14
	Pb without inoculation		6.66 ± 0.12
	Pb + *P. fluorescens* 21		6.58 ± 0.10

Values are means ± standard deviations

Soil Fractions' Pb and Soil pH

In Experiments 2 and 3, in the stooling and heading stages in the presence of both *P. fluorescens* strains, the concentrations of lead in the soil fraction extracted with ammonium acetate were found to increase 1.5–1.6-fold relative to the corresponding uninoculated variants (Figure **1**, Table **3**). Inoculation with *Pseudomonas* increased the Pb concentration in this soil fraction in the stooling stage from 24% to 35%; in the heading stage, from 12% to 18–20% of the applied heavy metal rate. Determination of the fractional composition of Pb compounds in soil by sequential extractions in the heading stage in Experiment 3 showed that the Pb content in the fraction bound to organic matter extractable with potassium pyrophosphate was up to 1.5 times greater at the application of both *P. fluorescens* strains (Figure **2**, Table **4**). At the same time, the metal content in the fraction bound to Fe oxides and hydroxides was found to be decreased; that in the residual fraction bound to clay minerals, to be slightly (at the application of *P. fluorescens* 20) or significantly (*P. fluorescens* 21) increased.

Application of the bacteria did not affect the pH of lead-contaminated soil in the first half of the growing season (Table **5**).

DISCUSSION

In the reported experiments, inoculation with *Pseudomonas* bacteria promoting the growth and increasing the biomass of various agricultural crops protected barley plants from the inhibitory effect of lead in soil artificially contaminated with water-soluble compounds of this heavy metal. In the presence of bacteria, the concentration of lead in shoots and roots noticeably decreased in the first half of the growing season (up to the heading stage), when plants are the most sensitive to unfavourable soil conditions, increased concentrations of heavy metals including. Earlier studies found lead-resistant pseudomonads to promote the growth of wheat seedlings under lead stress conditions (Hasnain et al. 1993). Some plant growth-promoting bacteria were reported (Höflich and Metz 1997) to be capable of stimulating the growth of maize and the uptake of lead by plants grown on heavy metal-polluted soil. In contrast, inoculation with growth-promoting *Kluyvera ascorbata* bacteria was noted not to affect the accumulation of lead in plants under heavy metal-induced stress conditions (Burd et al. 2000). According to another work (Wu et al. 2006) application of the bacterium *Azotobacter chroococcum* HKN-5 stimulated the growth of brown mustard *Brassica juncea*, and the use of two strains of the bacterium *Bacillus megaterium* protected plants from the toxicity of Pb–Zn mining tails, but had a minor effect on the concentrations of metals in plant tissues. On the contrary, yet another work (Malcová et al. 2003) established a decrease in the concentration of Pb in shoots and roots of *Zea mays* L. and *Agrostis capillaris* L. that were associated with *Glomus intraradices*. In our studies, inoculation with the most efficient bacterium *P. fluorescens* 21 not only significantly reduced the Pb content in shoots and roots in the first half of the growing season but also stimulated plant growth starting from the heading stage and increased grain weight 1.5-fold under contaminated conditions. Besides, in the heading and maturity stages plants inoculated with strain *P. fluorescens* 21 in Pb-treated soil had the same weight, of grain including, as in the corresponding uninoculated and uncontaminated variants. Thus, application of this strain eliminated Pb toxicity and enhanced the resistance of plants to stress caused by heavy metal.

It is known that the pH of soil has a direct impact on the mobility of metals. Concentrations of divalent cations in soil solution and their uptake by plants usually decrease at high pH values (Khan et al. 2009). The present observations showed that the decrease of Pb concentration in inoculated plants proceeded without changes in the reaction of the soil medium and, therefore, other mechanisms were responsible for the process. Fluorescent species of *Pseudomonas* are ubiquitous soil inhabitants and produce high amounts of siderophores – chelating low molecular agents characterized by extremely high affinities for Fe (III) (Neilands and Leong 1986; Loper and Buyer 1991). Siderophores are able to sequester many other metals in addition to iron, heavy metals such as lead including, but with a lower affinity (Braud et al. 2010). A siderophore produced by the bacterium *Pseudomonas stutzeri* str. KC as a thiol-containing compound reacted with toxic heavy metals to precipitate them, including Pb as sulfides, to transform metals into insoluble and less toxic state (Zawadzka et al. 2009).

In the present experiments, at the initial stages of plant development (up to the heading stage), *Pseudomonas* bacteria significantly increased the content of Pb in the soil fraction extracted with ammonium acetate to reach one-third of the metal introduced into soil as $Pb(NO_3)_2$. Inoculation with bacteria also increased the content of metal in the organic fraction extracted with potassium pyrophosphate. A decrease in the concentration of Pb in plant tissues at the initial stages of the ontogenesis after the inoculation with *Pseudomonas* bacteria corresponded to a 1.5-fold increase in the content of metal in the soil fraction extracted with ammonium acetate as well as in the organic fraction with respect to the uninoculated variants. It is quite likely that the Pb resistance of plants inoculated with *Pseudomonas*, especially with *P. fluorescens* 21, could be due to the binding of Pb in the above-mentioned soil fractions and, as a consequence, a decrease in the uptake of heavy metal by shoots and roots in the early periods of plant ontogenesis.

In soils, lead accumulates mainly in compounds bound to organic matter and iron minerals (Tessier et al. 1979; Plekhanova and Bambusheva 2010). Ammonium acetate buffer (pH 4.8) extracts from soil a fraction of heavy metals such as lead that is retained by covalent or coordination forces on clay minerals, on the surfaces of amorphous precipitates of aluminosilicates, silica, oxides of Fe, Al and Mn, and on crystallized precipitates of oxides, carbonates and phosphates (Plekhanova and Bambusheva 2010). Lead was adsorbed as a metal–siderophore complex on the surface of kaolinite clay in the presence of trihydroxamate siderophore desferrioxamine-B under certain pH conditions (Hepinstall et al. 2005; Mishra et al. 2007), whereby Pb^{2+} bridged the clay surface and the siderophore molecule (Mishra et al. 2007). An increase in the content of Pb in soil in the NH_4OAc-extracted fraction and in the $K_4P_2O_7$-extracted organic fraction at the inoculation with *Pseudomonas* could occur as a result of its binding by organic bacterial metabolites – siderophores – on the surface of various soil components. This is, probably, what led to a decrease of the mobility of metal.

On the other hand, a decrease of Pb concentrations in shoots and roots in the first half of the growing season, especially in the presence of *P. fluorescens* 21, could be caused by the dilutive effect due to increased biomass of inoculated plants (Lebeau et al. 2008) also observed in our experiments. In mature plants this phenomenon was not observed, probably, due to the accumulation of larger amounts of plant biomass as compared with shoots at the initial vegetation stages. The dilutive effect in grain at the application of *P. fluorescens* 21 bacteria was not observed, probably, also due to the low content of lead in grain. An insignificant Pb content in barley grain revealed in this study was due to the barrier functions of the root system of cereals, which reduced the translocation of heavy metal to shoots (Obroucheva et al. 2007). In all experiments (except Experiment 1, where this parameter was not determined) the roots contained on average by two orders of magnitude more lead that shoots, straw and grain. A decrease in the uptake of Pb by the root system of inoculated plants, even at an increased accumulation of metal in the roots is, probably, associated with an enhancement of the soil barrier functions with respect to lead, due to a decrease of its mobility in soil. Also, it is, probably, due to an enhancement of the physiological barrier functions (with respect to this element) at the soil–root interface.

Another possible mechanism of plant growth stimulation could be microbial production of IAA. It was shown to stimulate root elongation and is likely to play a key role in plant growth stimulation in the presence of heavy metals, too (Patten and Glick 2002). Rhizobacteria improved plant growth and development in heavy metal-contaminated soil by promoting root growth and branching (Jing et al. 2007). In (Gamalero et al. (2004), application of two *P. fluorescens* strains (92rk and P190r) was found to have a positive effect on the development of the root system in tomato and cucumber. Earlier laboratory studies have shown that *Pseudomonas* PGPR tested in this work produce a significant amount of IAA under batch cultivation (Olyunina and Shabaev 1996). In this work, the use of the PGPR *P. fluorescens* 21 also considerably enhanced root weight at almost all stages of plant development in Pb-treated soil.

Plant growth-promoting rhizobacteria can also reduce the toxicity of heavy metals by increasing the mobility and bioavailability of nutrients in the rhizosphere (Khan et al. 2009). Our previous pot experiments have found that a decrease of the negative effect of lead on barley plants inoculated with *Pseudomonas* PGPR is accompanied by a simultaneously increased nutrient uptake by plants (Shabayev and Ostroumov 2013). The protective action of *Pseudomonas* under lead stress conditions observed in the present experiments could be also associated with their improved mineral nutrition, an increased uptake of nutrients that are part of biologically important compounds in plants.

Conclusion

Thus, conducted pot experiments convincingly showed that application of *Pseudomonas* PGPR efficiently reduced the concentration of lead in shoots and roots of barley plants grown on gray forest soil (southern Moscow Region, Russia) artificially contaminated with water-soluble heavy metal compounds. Application of *Pseudomonas* increased the content of heavy metal in soil in the form of stable complexes and organic compounds in the first half of the growing season (up to the heading stage) without changing the reaction of the soil medium, probably, due to the binding of lead by organic bacterial metabolites, such as siderophores. Therefore, application of bacteria contributed to phytostabilization in the early periods of plant growth and protected young plants from the toxic action of lead, when they are most sensitive to unfavourable growth factors, increased concentrations of heavy metals including. The concentration of lead in shoots of inoculated plants could be also reduced due to growth dilution, an increase in shoot and root weight observed in these experiments. An increased tolerance of inoculated plants to the toxic action of lead is, presumably, also due to the enhanced development of the root system; a significant role in the process could be played by production of IAA by bacteria. A positive effect of *Pseudomonas* on barley plants under lead stress conditions is also due to an increased nutrient uptake by plants from soil.

Inoculation with the most efficient bacterium *P. fluorescens* 21 stimulated the growth of barley on Pb-contaminated soil starting from the heading stage to increase grain weight 1.5 times, to the level of the uninoculated control. These first results of pot experiments show that application of an efficient rhizobacterium eliminates the phytotoxicity of lead without the loss of agricultural crop productivity. From these studies, it can be concluded that the bacterium *P. fluorescens* 21 could be used to develop an inoculant to increase the productivity of a grain crop – barley – grown on Pb-contaminated soil. Application of rhizobacteria facilitating the phytostabilization of heavy metals in soil could be used in developing strategies of biological remediation of heavy metal-contaminated soils based on environmentally friendly approaches.

Acknowledgments

This work was supported by grant 12-04-01205-a of the Russian Foundation for Basic Research. The author would like to express his sincere gratitude to Mrs A.N. Maltseva for technical assistance throughout these studies.

References

Braud, A., Geoffroy, V., Hoegy, F., Mislin, G. L. A. & Schalk, I. J. (2010). Presence of the siderophores pyoverdine and pyochelin in the extracellular medium reduces toxic metal accumulation in *Pseudomonas aeruginosa* and increases bacterial metal tolerance. *Environmental Microbiology Reports*, *2*, 419-25.

Burd, G. I., Dixon, D. G. & Glick, B. R. (2000). Plant growth promoting bacterium that decreases heavy metal toxicity in plants. *Canadian Journal of Microbiology*, *46*, 237-45.

Gadd, G. M. (2000). Bioremedial potential of microbial mechanisms of metal mobilization and immobilization. *Current Opinion in Biotechnology*, *11*, 271-79.

Gamalero, E., Trotta, A., Massa, N., Copetta, A., Marinotti, M. G. & Berta, G. (2004). Impact of two fluorescent pseudomonads and an arbuscular mycorrhizal fungus on tomato plant growth, root architecture and P acquisition. *Mycorrhiza*, *14*, 185-92.

Gamalero, E., Lingua, G., Berta, G. & Glick, B. R. (2009). Beneficial role of plant growth promoting bacteria and arbuscular mycorrhizal fungi on plant responses to heavy metal stress. *Canadian Journal of Microbiology*, *55*, 501-14.

Hasnain, S., Yasmin, S. & Yasmin, A. (1993). The effects of lead resistant pseudomonads on the growth of *Triticum aestivum* seedlings under lead stress. *Environmental Pollution*, *81*, 179-84.

Hepinstall, S. E., Turner, B. F. & Maurice, P. A. (2005). Effects of siderophores on Pb and Cd adsorption to kaolinite. *Clays and Clay Minerals*, *53*, 557-63.

Höflich, G. & Metz, R. (1997). Interaction of plant–microorganism association in heavy metal containing soils from sewage farms. *Bodenkultur*, *48*, 238-47.

Islam, E. U., Yang, X. E., He, Z. L. & Mahmood, Q. (2007). Assessing potential dietary toxicity of heavy metals in selected vegetables and food crops. *Journal Zhejiang University Science B*, *8*, 1-13.

Jing, Y. D., He, Z. L. & Yang, X. E. (2007). Role of soil rhizobacteria in phytoremediation of heavy metal contaminated soils. *Journal Zhejiang University Science B*, *8*, 192-207.

Khan, M. S., Zaidi, A., Wani, P. A. & Oves, M. (2009). Role of plant growth promoting rhizobacteria in the remediation of metal contaminated soils. *Environmental Chemistry Letters*, *7*, 1-19.

Ladonin, D. V. (2006). The method for the determination of the fractional composition of heavy metal compounds in soils. In: Vorobyova, LF (Ed.) *Theory and practice of the chemical analysis of soils*. GEOS, Moscow, pp. 293-309 (in Russian).

Lebeau, T., Braud, A. & Jézéquel, K. (2008). Performance of bioaugmentation-assisted phytoextraction applied to metal contaminated soils: A review. *Environmental Pollution*, *153*, 497-522.

Loper, J. E. & Buyer, J. S. (1991). Siderophores in microbial interactions on plant surfaces. *Molecular Plant-Microbe Interactions*, *4*, 5-13.

Malcová, R., Vosátka, M. & Gryndler, M. (2003). Effects of inoculation with *Glomus intraradices* on lead uptake by *Zea mays* L. and *Agrostis capillaris* L. *Applied Soil Ecology*, *23*, 55-67.

McGrath, S. P., Chaudri, A. M. & Giller, K. E. (1995). Long-term effects of metals in sewage sludge on soil, microorganisms and plants. *Journal of Industrial Microbiology*, *14*, 94-104.

Mineev, V. G., Shabaev, V. P., Safrina, O. S. & Smolin, V. Yu. (1991). Effect of bacteria of the genus *Pseudomonas* on red beet yield and nitrogen uptake by plants. *Soviet Agricultural Sciences*, *9*, 25-9.

Minkina, T. M., Motuzova, G. V. & Nazarenko, O. G. (2009). Composition of heavy metal compounds in soils. Everest, Rostov-on-Don, (in Russian).

Mishra, B., Haack, E. A., Vasconcelos, I. F., Maurice, P. A. & Bunker, B. A. (2007). EXAFS determination of Pb and Cd speciation with siderophores and the metal/siderophore/kaolinite system. *In: Proceedings of AIP Conference*, *882*, pp. 196-198.

Neilands, J. B. & Leong, S. A. (1986). Siderophores in relation to plant growth and disease. *Annual Review of Plant Physiology*, *37*, 187-208.

Obroucheva, N. V., Ivanov, V. B., Sobotik, M., Bergmann, H., Antipova, O. V., Bystrova, E. I., Seregin, I. V. & Shpigun, L. K. (2001). Lead effects on cereal roots in terms of cell growth, root architecture and metal accumulation. In: Gašpaŕiková, O, Čiamporová, M, Mistrík, I, & Baluška F (Eds.) *Recent advances of plant root structure and function*. Kluwer Academic Press, The Netherlands, pp. 165-170.

Olyunina, L. N. & Shabaev, V. P. (1996). Production of indole-3-acetic acid during growth of the rhizosphere bacteria of the genus *Pseudomonas*. *Mikrobiologiya.*, *65*, 709-13.

Patten, C. L. & Glick, B. R. (2002). Role of *Pseudomonas putida* indoleacetic acid in development of the host plant root system. *Applied and Environmental Microbiology*, *68*, 3795-3801.

Plekhanova, I. O. & Bambusheva, V. A. (2010). Extraction methods for studying the fractional composition of heavy metals in soils and their comparative assessment. *Eurasian Soil Science*, *43*, 1004-10.

Shabaev, V. P. (2004). Optimization of mineral fertilizer doses by inoculating seeds with growth-stimulating rhizosphere bacteria when growing cereal crops. *Russian Agricultural Sciences*, *11*, 17-20.

Shabayev, V. P., Smolin, V. Yu. & Mudrik, V. A. (1996). Nitrogen fixation and CO_2 exchange in soybeans (*Glycine max* L.) inoculated with mixed cultures of different microorganisms. *Biology and Fertility of Soils*, *23*, 425-30.

Shabayev, V. P., Safrina, O. S. & Mudrik, V. A. (1998). The effect of rhizospheric bacterium *Pseudomonas fluorescens* 20 and VA mycorrhizal fungus *Glomus mosseae* on the yield and development of radish depending on nutrition conditions. *Agricultural Chemistry*, *2*, 126-33.

Shabayev, V. P. & Smolin, V. Yu. (2000). Sensitivity of winter wheat to inoculation with *Pseudomonas* bacteria in gray forest soil. *Eurasian Soil Science*, *33*, 434-40.

Shabayev, V. P. & Smolin, V. Yu. (2002). The response of fodder beet to inoculation with *Pseudomonas fluorescens* 20 on gray forest soil as conditioned by mineral nutrition level. *Eurasian Soil Science*, *35*, 634-42.

Shabayev, V. P. & Voronina, L. P. (2007). Grain yield and quality of winter wheat inoculated with a mixed culture of *Pseudomonas* bacteria against the background of increasing nitrogen fertilizer rates. *Russian Agricultural Sciences*, *33*, 311-13.

Shabayev, V. P. & Ostroumov, V. E. (2013). Application of growth-promoting rhizosphere bacterium for remediation of Pb-contaminated gray forest soil. *Russian Agricultural Sciences*, 5–6, 443-46.

Sharma, P. & Dubey, R. S. (2005). Lead toxicity in plants. *Brazilian Journal of Plant Physiology*, *17*, 35-52.

Tessier, A., Campbell, P. G. C. & Bisson, M. (1979). Sequential extraction procedure for the speciation of particulate trace metals. *Analytical Chemistry*, *51*, 844-51.

Wu, S. C., Cheung, K. C., Luo, Y. M. & Wong, M. H. (2006). Effects of inoculation of plant-growing rhizobacteria on metal uptake by *Brassica juncea*. *Environmental Pollution*, *140*, 124-35.

Zawadzka, A. M., Paszczynski, A. J. & Crawford, R. L. (2009). Transformations of toxic metals and metalloids by *Pseudomonas stutzeri* strain KC and its siderophore pyridine-2,6-bis (thiocarboxylic acid). In: Singh, A, Ramesh, CK & Ward, OP (Eds.) *Advances in Applied Bioremediation*, *Soil Biology*, *17*, Springer-Verlag, Berlin, Heidelberg, pp. 221-238.

Part III. Risk Estimates and Management in Impacted Polar Ecosystems Based On Ecological and Biogeochemical Cycling

In: Ecological and Biogeochemical Cycling ... ISBN: 978-1-53612-081-3
Editor: Vladimir N. Bashkin

Chapter 12

RISK MANAGEMENT OF THE ENVIRONMENTAL POLLUTION BY METHANOL AS THE MAIN INHIBITOR OF HYDRATE FORMATION IN THE GAS INDUSTRY OF POLAR REGIONS

Rauf V. Galiulin[1,*], Vladimir N. Bashkin[2,3] and Rosa A. Galiulina[1]

[1]Institute of Basic Biological Problems of Russian Academy of Sciences, Pushchino, Russian Federation
[2]Gazprom VNIIGAZ LLC, Razvilka, Russian Federation
[3]Institute of Physicochemical and Biological Problems of Soil Science of Russian Academy of Sciences, Pushchino, Russian Federation

---

*Address correspondence to this author at the Institute of Basic Biological Problems of Russian Academy of Sciences: Pushchino, Moscow Region, 142290, Russian Federation; Tel: +7 4967 33-14-53; Fax: +7 4967 33-05-32, E-mail: rauf-galiulin@rambler.ru..

Abstract

The examples and risks of the environmental pollution by methanol applied as the main inhibitor of hydrate formation in the gas industry in the Polar regions are reviewed, as well as toxic effects of methanol on human health are considered. Hygienic standards for control of environmental pollution by methanol are submitted. Various methods of utilization and cleaning of sewage and soils containing methanol are characterized (thermal, underground burial, physicochemical, catalytic and microbiological).

Keywords: gas industry, methanol, hydrate formation inhibitor, environmental pollution, risk of toxic action on human health, control of environmental pollution, methods of utilization and cleaning

Introduction

Methanol (CH_3OH) is widely used in extraction of natural gas as hydrate formation inhibitor, i.e., for fight against such undesirable phenomenon as formation under the defined thermobaric conditions from water and low-molecular gases, so-called gas hydrates in the form of solid crystal compounds (Bashkin et al. 2014). Methanol is known to belong to the class of thermodynamic inhibitors of hydrate formation. So it may decrease water activity in aquatic solutions and prevent the formation of the methane hydrates. So, pumping of methanol in a bottomhole zone of the well of gaseous-hydrate fields causes not only degradation of gas hydrates on a well face, but also improves filtration characteristics of a bottomhole zone, i.e., of the site of layer adjoining a well tube. Besides, high adsorptive ability of methanol is used for removal of water after hydrostatic tests of gas pipelines, and also in low-temperature processes of cleaning of natural gas from carbon dioxide (CO_2), hydrogen sulfide (H_2S) and other sulfur-containing organic compounds.

General use of methanol, especially at the gas enterprises of the Far North, is caused by a variety of reasons among which (Grunvald 2007):

1. the relatively its low cost in comparison with other inhibitors of hydrate formation (glycols, surfactants, water-soluble polymeric compositions);
2. the highest anti-hydrate activity among the known inhibitors remaining even at low temperatures;
3. the very low temperature of freezing of the concentrated solutions of methanol and exclusively small their viscosity even at a temperature is lower - 50°C.

The main objective of this work consisted in the analysis, systematization and generalization of information concerning risks of environmental pollution by methanol, its toxic action on human and control of environment pollution by methanol and also the methods of utilization and cleaning of sewage and soils containing methanol.

EXAMPLES OF ENVIRONMENTAL POLLUTION BY METHANOL

Meanwhile the risk of the environment contamination by methanol can be shown in the form of emergency emissions or spills in case of its production, transportation and use of this substance. At the same time the number of the emergency emissions or spills of methanol or industrial sewage containing this substance can't be planned, and to avoid them for 100% impracticably.

So, recently, in the Sverdlovsk region at the railway station there was a leakage of a significant amount of methanol (850 l) from the tank on the railway track (Dmitrevskaya et al. 2014). A series of the incidents connected with high pollution of atmospheric air by methanol i.e., up to the 10, 15 and 22 of maximum permissible concentration were registered in the Tula region (Dmitrevskaya et al. 2016; 2016a; 2016b). Extremely high pollution by methanol of river water up to the 58 of maximum permissible concentration was established in the Vologda region (Dmitrevskaya et al. 2016b). In one of the cities of the Kemerovo region in water of wells on territory of the chemical enterprises the methanol in concentration exceeding its maximum permissible concentration was found (Elpiner 2015). The methanol to number of priority pollutants of drinking water sources requiring control was carried in one of the cities of the Arkhangelsk region (Unguryanu 2011).

Risk of Environmental Pollution by Methanol

The greatest risk of the environment contamination is represented the methanol transportation on the gas enterprises. It is known that the transport scheme of providing the gas enterprises with methanol existing now on the Far North rather difficult and includes several stages, namely (Grunvald 2007):

1. filling of methanol in railway cisterns on manufacturing plant and their transportation on head base;
2. methanol decantation from railway cisterns in stationary capacities for storage;
3. preparation of methanol for its use by addition of dye-stuff or odorant;
4. methanol decantation from stationary capacities in car cisterns and their transportation to base of methanol at the gas enterprise,
5. methanol decantation from a car cistern in the stationary capacity,
6. methanol decantation from stationary capacity in other car cisterns and their transportation on concrete objects of consumption.

The high risk, in particular, for the water environment is represented by transportation of methanol on cargo ships directly on sea platforms of gas production in the region of the Far North when emergency spills of methanol can affect safety of valuable breeds of fishes in reservoirs of the highest fishery category (Iunusov et al. 2008, Dolinsky 2009). The example of transportation of methanol the existing now delivering on the Ob River and the Taz Bay is given in works (Iunusov et al. 2008, Dolinsky 2009).

Alternative to now applied transportation of methanol on a long distance to the place of use when quite often there are transport accidents is practical implementation so called the low-tonnage production of this substance directly around gas production in the form of the mini-plants (Iunusov et al. 2008, Dolinsky 2009; Ladygin et al. 2014). At the same time as initial raw materials for production of methanol can serve the natural gas of the field, by way obtaining through method of steam conversion from natural gas so called synthesis-gas (mixture of carbon monoxide and hydrogen) and its subsequent catalytic transformation into methanol on the copper-zinc oxidic catalyst:

$$CH_4 + H_2O \rightarrow CO + 3H_2; CO + 2H_2 \rightarrow CH_3OH$$

It is remarkable the fact that main part of natural gas is constituted by methane (CH_4) - from 70 to 98%.

The mini-plant of a certain capacity on the gas field mounted in quickly reproduced modular building with the simplified technological scheme, allows to reduce or to completely stop transportation of methanol on a long distance from places of its production, and, therefore, to exclude an environment contamination risk methanol (Ladygin et al. 2014).

Risk of Toxic Effect of Methanol on Human

The methanol is strong, mainly nervous and vascular poison with sharply expressed cumulative effect, i.e., toxic action as a result of its accumulation in a human organism at multiple enterings (Bashkin et al.

2014). The methanol at oral entering in a human organism causes a circulatory collapse, i.e., the acute vascular failure, which is followed by sharp falling of blood pressure. Special toxicity of methanol is connected with formation from it in an organism of formaldehyde (HCOH) and formic acid (HCOOH). At any way of intake of methanol in organism, the damages of an optic nerve and eye retina are typical which noted both at sharp and at chronic intoxications. Methanol vapors strongly irritate mucous membranes of eyes and airways. Entering of a methanol in a stomach is dangerous in quantity even 5-10 ml, and a lethal dose is 30 ml. Poisoning symptoms (nausea, vomiting) can come as soon after substance entering in organism, and in several hours, on next day or is even later. In hard cases are observed sharp cyanosis, the deep and complicated respiration, cramps, the low speeded-up pulse, lack of reaction of pupils, and death occurs from respiratory standstill. The victims who are in consciousness complain of a headache, the most severe all body pains and in a stomach, flashing before eyes and a vision ambiguity. Not disappearing expansion of pupils specifies a possibility of a recurrence or permanent disorder of sight. Functional inferiority of a liver doesn't disappear with approach of clinical recovery, which proceeds slowly.

Early symptoms of chronic intoxication by methanol are shown in the form of concentric narrowing of limits of the color sight increasing over time and an atrophy of an optic nerve, i.e., reduction of its sizes, followed by violation or the termination of function and hypostasis. Persons with chronic intoxication of methanol under production conditions have a change of protein-forming function of a liver. At small concentrations of methanol poisoning develops gradually and is characterized by irritation of mucous membranes, frequent respiratory diseases, headaches, ringing in one's ears, neuritis and disorders of sight. Organism poisoning at entering of methanol on skin usually occurs at simultaneous inhalation of its vapors.

Meanwhile for an exception features for the wrong use of methanol on production in it are added by odorants - ethylmerkaptan (C_2H_5SH, 1:1000) and kerosene (1:100) or dark dye-stuff (2.5:1000).

Control of Environmental Pollution by Methanol

Control of methanol content in the environment is carried out according to hygienic standards in the form of its maximum permissible concentration in air and water, and on the human in production conditions in the form of maximum permissible level of methanol on skin of his hands, Table 1 (Bashkin et al. 2014).

Table 1. Hygienic standards of methanol in various environments and for human

Hygienic standard	Environment and person	Value
Maximum permissible concentration	In air of a working zone	5 mg/m^3
	Maximum single in air of the inhabited places	1 mg/m^3
	Average daily in air of the inhabited places	0.5 mg/m^3
	In water of water objects	3 mg/l
Maximum permissible level	On skin of hands	0.02 mg/cm^2

Here various hygienic standards are understood as the following:

1) maximum permissible concentration of substance in air of a working zone is concentration which during all length of service shouldn't lead to a disease or a deviation in the state of health;
2) maximum permissible concentration of substance as maximum single is a concentration in air of the inhabited places which at inhalation within 30 minutes shouldn't cause reflex reactions in a human body;
3) maximum permissible concentration of substance as average daily is a concentration in air of the inhabited places which shouldn't make on the person negative impact at vaguely long inhalation;

4) maximum permissible concentration of substance in water of a reservoir is a concentration which shouldn't exert negative impact on a human organism and shouldn't worsen hygienic conditions of water use;
5) maximum permissible level of substance on skin of hands is a concentration which during all length of service shouldn't lead to a disease or a deviation in the state of health.

The researches (Boyko et al. 2016) can be as real example of control of methanol content in the air environment, carried out in production premises of the Astrakhan gas processing plant. So it was established that the maximum contents of methanol were lower or slightly exceeded the maximum permissible concentration equal of 5 mg/m^3 established for air of a working zone.

Methods of Utilization and Cleaning of Sewage and Soil Containing Methanol

It is known that the sewage formed at enterprises of the gas industry contains some other specific components (hydrocarbons, phenols, glycols and hydrogen sulfide) along with methanol (Akopova et al. 2003). Utilization of this sort of sewage by a thermal method, i.e., burning on gas-torch installations isn't ecologically safe as dangerous products of combustion of components of sewage arrive in atmospheric air, then settle on soil and open water objects.

Other method of utilization of sewage the underground burial of sewage is carried out by their pumping in the deep, reliably isolated water-bearing horizons which aren't containing fresh, balneological, mineral and thermal waters. At the same time underground burial of sewage to the area of a depression funnel in water-pressure system of the developed natural gas field can be carried out at impossibility of sewage cleaning from methanol and other components to the required maximum permissible concentration.

It is known the physicochemical method of sewage cleaning with the content of methanol in several its maximum permissible concentration value which makes 1 mg/l, with use of ultra-violet radiation exilamps (gas-discharge lamps) in the presence of nitric acid (HNO_3) as a strong oxidizer (Medvedev et al. 2005). At the same time under the influence of ultraviolet radiation happens photolysis of water and nitric acid with formation of high-reactive radicals - $^{\bullet}OH$, $^{\bullet}H$, $NO_2^{\bullet}$ and $^{\bullet}NO$ which react further with methanol with formation of end products of CO_2, H_2O and NH_3. In the conditions of *in vitro* of an experiment it was set that in the sewage containing methanol with adding of nitric acid (at ratio of CH_3OH:HNO_3, 10:1) in 16 minutes, under the influence of an ultraviolet radiation with wavelength $\lambda = 172$ nanometers, and with use of Xe_2-exilamp, concentration of substance in water decreased by 13 times, and when using an ultraviolet radiation with wavelength $\lambda = 222$ nanometers, and with use of KrCl-exilamp, decreased by 23 times.

In work (Brenchugina et al. 2007) the technological scheme of extraction of methanol from sewage of gas-condensate fields consisting in regeneration of this substance by rectification with the subsequent deep catalytic oxidation of its residual quantities to maximum permissible concentration is offered. At the same time 100% oxidation of methanol in concentration up to 1.5% are reached when using the copper-chromium-magnesium and chromium-magnesium catalyst on the carrier from oxide of aluminum (Al_2O_3). At the same time duration of contact of the sewage containing methanol with oxidation carried out at a temperature 450°C has made 0,9 seconds. Meanwhile researches (Sharkina et al. 2012) have shown 100% cleaning of sewage from methanol on a copper-zinc-chromium catalyst at 250°C and with initial content of substance of 5%.

In other method of cleaning not only sewage from methanol, but also soil are used microorganisms. So, in works (Murzakov et al. 2005, 2006) practical recommendations about cleaning of the specified media by means of biopreparations in the form of the dried-up active biomass the methylotroph bacteria (*Acinetobacter calcoaceticus* and *Methylomonas methanica*) selected from lake water and soil are made. Cleaning of the polluted media from methanol happens by microbiological transformation (oxidation) of this substance through formaldehyde and formic acid to carbon dioxide and water:

$$CH_3OH \rightarrow HCOH \rightarrow HCOOH \rightarrow CO_2 + H_2O$$

At the same time cleaning of sewage can be carried out directly in the special ponds-stores equipped with system of compressors for forcing of air in the volume of the cleaning water and its simultaneous processing by the biopreparation. So, in the conditions of a fermenter it has been established that at 1% the content of methanol in sewage its full cleaning of substance happened in 22 hours, at 2% the content of methanol - in 36 hours.

Meanwhile, for decrease in migration risk of methanol from the polluted soil in surface and underground waters there is a need of its expeditious cleaning which is also carried out by means of the above-stated biopreparations (Murzakov et al. 2005). So, at superficial (0-5 cm) and subsurface (5-30 cm) pollution of the soil by methanol it is processed specially prepared biopreparation suspension (in solution of mineral fertilizers). At the same time before and after of processing by a biopreparation the upper layers of a soil profile subject to loosening. In case of depth pollution of a soil profile by methanol (up to 100 cm), its layer is completely excavated and stored in the form of a clamp on specially prepared platform with the waterproof base and system of the punched pipes passing through thickness of a clamp and providing intensive aeration by means of compressors. The clamp process by biopreparation, periodically subject to loosening and after cleaning return an excavated layer to the place of dredging. For cleaning of the lower layers of a soil profile lay wells on all depth of pollution up to a mirror of ground waters in which through the punched pipes pump over suspension of a biopreparation and air.

CONCLUSION

Thus, using methanol as the main inhibitor of hydrate formation in the gas industry in the Polar regions can be accompanied by the contamination risks of the environment. It results due to of emergency emissions or spills in case of methanol production, transportation and application. Need of control of the environmental pollution by methanol is connected with risk of its toxic action on human and for this purpose use hygienic standard of substance in the form of maximum permissible concentration and maximum permissible level. Methods of utilization and cleaning of sewage and soil containing methanol are various (thermal, underground burial, physicochemical, catalytic and microbiological) and which choice for practical application, first of all, will be determined by the most effective environment protection and health of human.

CONFLICT OF INTEREST

The authors confirm that this article content has no conflict of interest.

ACKNOWLEDGMENTS

The authors express their deep appreciation and gratitude for their affiliation institutions, namely Institute of Basic Biological Problems of Russian Academy of Sciences and also Institute of Physicochemical and Biological Problems of Soil Science of Russian Academy of Sciences for the support during research and development.

REFERENCES

Akopova, G. S., Ilchenko, V. P. & Popadko, N. V. (2003). Production sewage of the gas industry: formation sources, composition, cleaning and utilization. *Gas Industry*, 6, 76-8.

Bashkin, V., Galiulin, R., Galiulina, R. & Grunvald, A. (2014). CH_3OH. *Neftegaz. RU*, 11-12, 84-6.

Boyko, O. B., Akhmineeva, A. K., Boyko, V. I. & Gudinskaya, N. I. (2016). Influence of the Astrakhan gas processing plant on air pollution of production premises and territory. *Hygiene and Sanitation*, 2 (95), 167-171.

Brenchugina, M. V., Buynovsky, A. S., Ismagilov, Z. R. & Kuznetsov, V. V. (2007). Development of technology of cleaning production waters of gas-condensate fields from methanol. *News of Tomsk Polytechnical University*, 3 (311), 64-8.

Dmitrevskaya, E. S., Krasilnikova, T. A. & Markova, O. A. (2014). About the environment pollution and radiation situation on the territory of the Russian Federation in March, 2014. *Meteorology and Hydrology*, 6, 103-110.

Dmitrevskaya, E. S., Krasilnikova, T. A. & Markova, O. A. (2016). About the environment pollution and radiation situation on the territory of the Russian Federation in May, 2016. *Meteorology and Hydrology*, 8, 100-6.

Dmitrevskaya, E. S., Krasilnikova, T. A. & Markova, O. A. (2016a). About the environment pollution and radiation situation on the territory of the Russian Federation in June, 2016. *Meteorology and Hydrology*, 9, 97-104.

Dmitrevskaya, E. S., Krasilnikova, T. A. & Markova, O. A. (2016b). About the environment pollution and radiation situation on the territory of the Russian Federation in July, 2016. *Meteorology and Hydrology*, 10, 103-110.

Dolinsky, S. E. (2009). Installations on methanol production behind the Polar Circle. *Gas Chemistry*, 4, 14-21.

Elpiner, L. I. (2015). Modern medico-ecological aspects of the doctrine about underground waters. *Hygiene and Sanitation*, 6 (94), 39-46.

Grunvald, A. V. (2007). Growth of methanol consumption in the gas industry of the Russia and the geoenvironmental risks arising at its use as hydrate formation inhibitor. *Oilgas Business*, p. 25. http://www.ogbus.ru.

Iunusov, R. R., Shevkunov, S. N., Dedovets, S. A., Ushakov, S. N., Lyats, K. G. & Samoylov, A. P. (2008). Low-tonnage installations on production of methanol in the gas regions of the Far North. *Gas Chemistry*, 1, 58-61.

Ladygin, K. V., Tsukerman, M. Y. & Stompel, S. I. (2014). Methanol in gas production: decrease of environmental risks. *Ecology of Industry*, 4, 47-9.

Medvedev, I. V., Polygalov, I. I., Erofeev, V. I., Erofeev, M. V., Sosnin, E. A., Tarasenko, V. F. & Istomin, V. A. (2005). Radiation of methanol solutions by Xe_2- and KrCl-exilamps of barrier category. *Gas industry*, 2, 63-5.

Murzakov B. G., Akopova G. S. & Markina P. A. (2005). Cleaning of methanol-containing waters by means of biological preparations. *Gas Industry*, 2005, 12, 58-60.

Murzakov, B. G., Akopova, G. S. & Markina, P. A. (2006). Selection of the methylotroph bacteria from microbiocenosis of methanol-containing of waters. *Gas Industry*, 3, 83-5.

Sharkina, V. I., Seregina, L. K., Shchankina, V. G., Falkevich, G. S. & Rostanin N. N. (2012). Cleaning of water-methanol fraction from methanol on the industrial NTK-4 catalyst. *Catalysis in the Industry*, 1, 61-4.

Unguryanu, T. N. (2011). Risk for population health at complex effect of the substances polluting drinking water. *Ecology of Human*, 3, 14-20.

In: Ecological and Biogeochemical Cycling … ISBN: 978-1-53612-081-3
Editor: Vladimir N. Bashkin

Chapter 13

RISK MANAGEMENT OF THE ENVIRONMENTAL POLLUTION DURING OIL PRODUCTION AND TRANSPORT

Rauf V. Galiulin[1,*], Vladimir N. Bashkin[2,3] and Rosa A. Galiulina[1]

[1]Institute of Basic Biological Problems of Russian Academy of Sciences, Pushchino, Russian Federation
[2]Institute of Physicochemical and Biological Problems of Soil Science of Russian Academy of Sciences, Pushchino, Russian Federation
[3]Gazprom VNIIGAZ LLC, Razvilka, Russian Federation

ABSTRACT

The analysis, systematization and generalization of information concerning a problem of environmental protection against oil contamination are carried out. Examples of the most considerable emergency oil floods to soil and surface waters during its production and

* Address correspondence to this author at the Institute of Basic Biological Problems of Russian Academy of Sciences: Pushchino, Moscow Region, 142290, Russian Federation; Tel: +7 4967 33-14-53; Fax: +7 4967 33-05-32, E-mail: rauf-galiulin@rambler.ru.

transport are presented. The measures for a remediation of the surface waters polluted by oil are described. Special attention is paid to remediation of oil polluted soils, for preventing both soil and ground water comtamination since oil may migrate through soil profile. The essence of this remediation consists in soil cleaning by means biocompost for the accelerated microbiological degradation of oil hydrocarbons. Researches showed that entering various doses of biocompost (50 and 100 g/kg) into the soil reduce time of almost full degradation of oil hydrocarbons at concentration in 50 g/kg by 1.8-5.4 times, at concentration of 100 g/kg – 1.7-7.2 times.

Keywords: oil, contamination, soil, surface waters, ground waters, protection, remediation, biocompost, oil hydrocarbons degradation

INTRODUCTION

It is known that the most serious pollution of soil, surface and underground waters by such fossil fuel as oil happens at its emergency flood. At the same time oil, coming in particularly to various water objects (lakes, rivers and seas), forms the film floating on water, the dissolved or emulsified forms, and its heavy fractions settle on a bottom and are adsorbed by bottom sediments as a source of secondary pollution of water mass (Sverlova & Voronina 2001). The last occurs at a stirring up of surface layers of bottom sediments at wind influence, sharp increase in speed of a stream, dredging, i.e., withdrawal of bottom sediments during the dredging works. Oil in the water environment makes direct or indirect negative impact on the organisms (hydrobionts) living in its. So, at a flood of 1 t of oil can be formed the film capable to cover up to 12 km^2 of a water mirror that leads to violation of the vital exchange processes between an atmosphere and a hydrosphere, to temperature increase of a surface layer of water, deterioration in gas exchange and eventually, to death of hydrobionts (Alekseeva & Masaeva 2008).

Meanwhile, the main source of intake of oil in underground waters is the soil, which becomes polluted at the accidents arising by search and investigation, production and transportation of oil. So, in the course of drilling and operation of wells tightness of the underground water-bearing horizons is disturbed, their hydrodynamic and hydrochemical modes change. Consequently, oil can migrate to the waters, which are already used for the drinking purposes, or subjected to use in the future (Zholdakova & Belyaeva 2015). Migration of oil from a soil in underground waters comes on pores of the horizons (on emptiness, cracks and cavities) composing a soil profile. Oil is besieged in zones of a capillary border and seasonal raising of underground waters, and also spreads on their surface. Furthermore, time achievement time by an oil stream of a mirror of underground waters is considerably reduced in the conditions of the so-called physically disturbed soil i.e., deprived of a vegetable cover and the top organogenic layer and presented by the underlying mineral horizons which have come to a day surface, and, therefore, the oil which is not protected from migration on a soil profile. Similar physical violation of the soil usually happens when constructing facilities on production and transportation of oil, in particular, on boring platforms and on routes of the roads and pipelines laid at the device of wells. In the absence of the top, most biologically active layer of soil, soil self-cleaning from oil, usually carried out by various hydrocarbons-oxidizing microorganisms (bacteria, yeast and fungi) is practically not observed (Koronelly 1996). Therefore, high risk of migration of oil into underground waters on a soil profile arises.

The purpose of this work is the analysis, systematization and generalization of information concerning a problem of protection of soil, surface and underground waters against pollution by oil. The logic of the specified problem demanded a statement of this information in the following sequence:

1) Emergency oil spills in soil and surface waters;
2) Remediation of surface waters polluted by oil;
3) Risk of oil migration in underground waters from polluted soil on its profile;
4) Remediation of polluted soil as source of oil entering in underground waters.

Emergency Oil Spills in Soil and Surface Waters

Recently the unauthorized (criminal) oil pipeline insert in the Saratov region resulted in pollution of water by oil of one of inflows of the Volga River with the content of substance in 40 maximum permissible concentrations (MPC) (Dmitrevskaya *et al.*, 2013). For the same reason the inflow of Volkhov River in the Leningrad region up to 82 MPC and the lake Zaton in the Samara region to 16 MPC have been polluted by oil (Dmitrevskaya et al., 2015; 2015c). In the Perm region there was a rush of a string of the infield oil pipeline which lead to pollution of the soil by oil on the square ~ 1200 m^2 (Dmitrevskaya et al., 2015). In the Krasnodar region owing to a depressurization of the trunk pipeline there was an oil spill on soil which got to the river of Tuapse and from its to the Black Sea (Dmitrevskaya et al., 2015a). For the similar reason there was an oil leak from the infield pipeline in the Orenburg region that has led to pollution of agricultural use land, water of a local stream and pond up to 18 MPC (Dmitrevskaya et al., 2015b). For the same region, in one case owing to oil spill from an oil-gathering collector there was a pollution of the soil and water of a stream up to 46 MPC, and in other case, because of depressurization of the oil pipeline water of a stream has become polluted to 23 MPC (Dmitrevskaya et al., 2016). In the Republic of Komi degreasing of oil wells has led to emission of oil and pollution of river water by its up to 82 MPC.

As it is well known the quantity of similar emergency floods cannot be planned, and it is impossible to avoid them almost completely and therefore naturally a problem of a remediation of the surface water polluted by oil is arising.

Remediation of Surface Waters Polluted by Oil

To number of basic measures for a remediation of the surface waters polluted by oil the following is carried:

1) Installation the booms (floating) obstacles having various modifications (fixed buoyancy, inflatable, tidal and emerging) made of the special fabric having the high durability and resistance to oil impact and employees for restriction of distribution of its film on a water surface and promoting its concentrated collection and also use of the sorbents (peat bertinate, i.e., the dehydrated peat, aerosil - pyrogenic silicon dioxide and butadienestyrene rubber) simplifying and accelerating the procedure of mechanical removal of oil from a water mirror (Kulikova & Dzerzhinskaya 2008);
2) Application for water purification of the hydrocarbons-oxidizing microorganisms preparations representing dried up by lyophilization (at a low temperature and in vacuum) biomass of active strains, mainly the bacteria using oil hydrocarbons as the only source of carbon and energy, and also as material for synthesis of structural components of a cell (Koronelly 1996). Nitrogen-phosphorus compounds for stimulation of growth of number of microorganisms, and also the neutral sorbent having buoyancy for keeping of bacteria on a surface of a hydrocarbonic film are the important part of these preparations. The mentioned preparations usually apply at the water pollution level by oil which is not allowing using water objects for the economic-drinking purposes, i.e., corresponding to an oil film on a water surface up to 1 mm thick.

Risk of Oil Migration in Undergrounds Waters from Polluted Soil on Its Profile

According to (Elpiner 2015), the share of underground waters in water supply of the population of the Russian Federation is quite considerable and makes 46%. And in this regard the importance is attached to protection of underground waters against pollution by oil in areas of its production and transportation, as there is a risk of migration of oil into underground waters from the polluted soil on its profile.

The real intake of oil from the polluted soil in underground waters is demonstrated by a number of the works performed in the conditions of *in situ*. So, in the experiment on the podzol illuvial-humus soil (The Middle Ob river) it has been established that the oil brought in soil surface at the dose of 22 l/m^2 in 1 year was frontally filtered up to the depth of 50 cm, that reaching a underground waters mirror causing their pollution (Kalachnikova et al., 1985).

On observations (Pikovsky et al., 1985), in the conditions of the earth barn intended for burial of spilled oil in a profile of the silty-gley soil (zone of humid subtropics) emergence of the powerful intra soil flow of oil moving to the place of unloading of underground waters was noted. At the same time the upper bound of oil flow was traced at a depth of 50-60 cm, and lower bound was closed with a underground waters mirror.

In other work (Maksimenko & Gersh 2006) content of oil around an earth barn with a total area of 3000 m^2 (The Krasnodar region) used for collection of the oil spreading around a well was determined. It was as a result established that oil pollution of a soil profile extended up to the depth of 120-130 cm.

Meanwhile, at intake of oil in underground waters, tastes of drinking water spoil already at content >0.1 mg/l. This value represents maximum permissible concentration in water, in particular, of the multisulphurous oil established on the so-called organoleptic limiting harm indicator characterizing change of a smell and taste of water in the presence of this substance. However water becomes soiled not only by oil, but also by the cancerogenic substance benz(α)pyrene ($C_{20}H_{12}$) which quantity depending on the oil field can fluctuate within 240-8050 mkg/kg that is fraught with grave consequences for the person health (Serkovskaya 2011). So, in researches (Belyakova et al., 2006), direct correlation between pollution of drinking water by benz(α)pyrene of an oil origin and incidence of gullet cancer in local population in the basin of the Urals River was established.

Remediation of Polluted Soil as Source of Oil Entering in Undergrounds Waters

Remediation of the polluted soil as a source of oil entering in underground waters includes soil cleaning by biological means containing hydrocarbons-oxidizing microorganisms. At the same time it is reasonable to carry out an efficiency evaluation of use of biological means originally in the conditions of *in vitro* that allows to model action of various factors (dose of substance, soil temperature and humidity) on degradation of oil hydrocarbons and to quickly obtain necessary information for an evidence-based remediation in the conditions of *in situ*.

So, in our work (Galiulin et al., 2010) efficiency of cleaning with physically disturbed and oil polluted soil by means of biocompost representing the fermented peat-manure mixture enriched with hydrocarbons-oxidizing microorganisms in number of 10^6 cell/g and also nutrients was estimated. Such quantity of microorganisms is considered sufficient for self-reproduction of their population as one of conditions of effective cleaning of polluted oil soil (Koronelly 1996). Physically disturbed soil representing the naked sandy loam lower illuvial horizon of the gray forest soil 50-90 cm thick served as an object of a research. Here the illuvial horizon is understood as the soil layer which is transition from the overlying horizons to parent rook. It should be noted that coloring of this soil is the most available, and first of all, the evident morphological feature characterizing its physical disturbance (Figure 1). Here for comparison of oil and biocompost colorings are presented. Meanwhile pollution of the soil was imitated by processing of its separate samples by oil in doses of 50 and 100 g/kg in which then brought biocompost in doses of 50 and 100 g/kg.

It should be noted that the peat which is a base part of biocompost slows down migration of oil in underground waters that is connected with its high sorption capacity reaching 10 g of oil on 1 g of dry peat and accelerates oil degradation process as number the hydrocarbons-oxidizing microorganisms in peat by 4-5 times exceeds a similar indicator for the soil (Alekseeva et al., 2000).

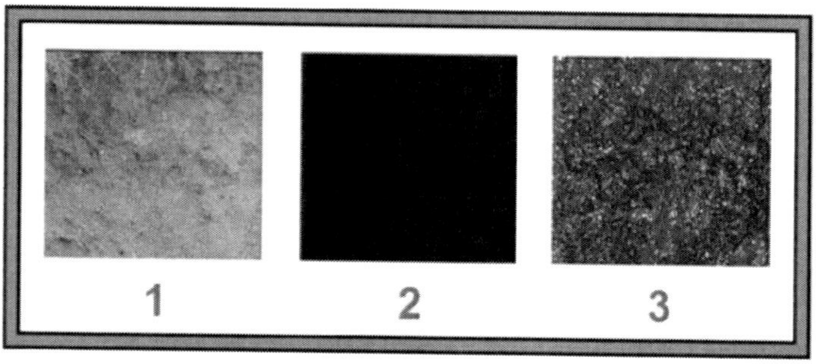

Figure 1. Coloring of the physically disturbed soil (1), oil (2) and biocompost (3).

Researches on degradation of oil hydrocarbons under the influence of biocompost have shown that when entering its various doses (50 and 100 g/kg) into the soil time of almost full degradation of hydrocarbons at concentration of oil in 50 g/kg is reduced rather control option by 1.8-5.4 times, and at concentration of 100 g/kg – 1.7-7.2 times (Table 1).

Table 1. Time of practically full degradation (T_{99}) of oil hydrocarbons in a physically disturbed and oil polluted soil under biocompost action

Variant	T_{99}, day	Variant	T_{99}, day
Oil, 50 g/kg, control	329	Oil, 100 g/kg, control	1151
Oil, 50 g/kg + biocompost, 50 g/kg	184	Oil, 100 g/kg + biocompost, 50 g/kg	658
Oil, 50 g/kg + biocompost, 100 g/kg	69	Oil, 100 g/kg + biocompost, 100 g/kg	288

Here calculation of time of almost full degradation of hydrocarbons, i.e., for 99% (T_{99}) was carried out on results of the analysis of content of hydrocarbons in dynamics, using exponential dependence: $y = e^{-kt}$, where y - the residual content of hydrocarbons for the period of t carried to initial y_0; e - basis of a natural logarithm; k - constant of speed of decomposition of hydrocarbons. The corresponding formula for calculation looks as: $T_{99} = ln\ 100/k$.

The contents of oil hydrocarbons noted in the course of the experiment a decrease under the influence of biocompost confirm the microbiological nature of degradation of these substances. The mechanism of observed process consists, first of all, in absorption of hydrocarbons by means of a hydrophobization of the cell wall of microorganisms realized through biosynthesis of specific compounds – lipophil glicol-, peptide- and peptideglicollipids (Koronelly 1996). In case of a direct contact of bacteria with a hydrocarbon film the latter penetrate into the cell by passive diffusion – gradual penetration into the cell wall – and reach enzymes on membranes, which are carrying out hydrocarbons degradation. Along with molecular-diffuse penetration of hydrocarbons through the surface of the cell wall, they can penetrate through special ultra-microscopic pores as well. Such channels filled with electron-dense (granular) substances were first found in yeasts. It is known that oxidation of oil hydrocarbons occur by majority of microorganisms happens to the help of so-called adaptive enzymes. The end products of oxidation of hydrocarbons are carbonic acid and water, however as intermediate products alcohols, organic acids and ester are found.

CONCLUSION

Thus, protection of soil, surface and underground waters against pollution by oil as the substance containing cancerogenic benz(α)pyrene, consists in their remediation. So, the boom obstacles are effective for remediation of the surface waters polluted as a result of emergency oil spills. These will restrict oil film moving on a water surface and promote oil collection. Other means for this purpose are the sorbents simplifying and accelerating the procedure of mechanical removal of oil from a water surface and also preparations of hydrocarbons-oxidizing microorganisms for cardinal water purification of in situ are used. Remediation of the polluted soil as main source of intake of oil in underground waters can be carried out only by its degradation by hydrocarbons-oxidizing microorganisms' preparations that the risk of pollution of these waters by this substance will be avoided.

Conflict of Interest

The authors confirm that this article content has no conflict of interest.

Acknowledgments

The authors express their deep appreciation and gratitude for their affiliation institutions, namely Institute of Basic Biological Problems of Russian Academy of Sciences and also Institute of Physicochemical and Biological Problems of Soil Sciences of Russian Academy of Sciences for the support during research and development.

References

Alekseeva, T. A. & Masaeva, L. M. (2008). Problems of protection of the water environment against pollution by oil and oil products. *Environment Protection in an Oil and Gas Complex*, 5, 21-4.

Alekseeva, T. P., Burmistrova, T. I., Tereshchenko, N. N., Stakhina, L. D. & Panova, I. I. (2000). Prospects of peat use for cleaning of the petropolluted soils. *Biotechnology*, 1, 58-64.

Belyakova, T. M., Dianova, T. M. & Trefilova, N. Ya. (2006). Geoecological studying of landscapes of a river basin the Ural in connection with population morbidity by cancer. *Problems of Biogeochemistry and Geochemical Ecology*, 1, 183-191.

Dmitrevskaya, E. S., Krasilnikova, T. A. & Markova, O. A. (2013). About pollution of environment and a radiation situation in the territory of the Russian Federation in May, 2013. *Meteorology and Hydrology*, 8, 101-7.

Dmitrevskaya, E. S., Krasilnikova, T. A. & Markova, O. A. (2015). About pollution of environment and a radiation situation in the territory of the Russian Federation in November, 2014. *Meteorology and Hydrology*, 2, 101-8.

Dmitrevskaya, E. S., Krasilnikova, T. A. & Markova, O. A. (2015a). About pollution of environment and a radiation situation in the territory of the Russian Federation in December, 2015. *Meteorology and Hydrology*, 3, 109-115.

Dmitrevskaya, E. S., Krasilnikova, T. A. & Markova, O. A. (2015b). About pollution of environment and a radiation situation in the territory of the Russian Federation in May, 2015. *Meteorology and Hydrology*, 8, pp. 103-111.

Dmitrevskaya, E. S., Krasilnikova, T. A. & Markova, O. A. (2015c). About pollution of environment and a radiation situation in the territory of the Russian Federation in July, 2015. *Meteorology and Hydrology*, 10, 101-8.

Dmitrevskaya, E. S., Krasilnikova, T. A. & Markova, O. A. (2016). About pollution of environment and a radiation situation in the territory of the Russian Federation in April, 2016. *Meteorology and Hydrology*, 7, 103-111.

Elpiner, L. I. (2015). Modern medico-ecological doctrines about underground waters. *Hygiene and Sanitation*, 6, 39-46.

Galiulin, R. V., Galiulina, R. A., Bashkin, V. N., Akopova, G. S., Listov, E. L. & Balakirev, I. V. (2010). Comparative assessment of gas condensate and oil hydrocarbons degradation in soil under effect of biological means. *Agrochemistry*, 10, 52-8.

Kalachnikova, I. G., Maslivets, T. A., Oborin, A. A., Ogloblina, A. I. & Pikovsky Iu. I. (1985). Transformation of oil in podsolic soils of Middle Ob river. *Migration of the Polluting Substances in Soils and Adjacent Environments*. Gidrometeoizdat Publishing House, Leningrad, pp. 74-80.

Koronelly, T. V. (1996). The principles and methods of an intensification of biological degradation of hydrocarbons in the environment (review). *Applied Biochemistry and Microbiology*, 6(32), 579-585.

Kulikova, I. Iu. & Dserzhinskaya, I. S. (2008). Modern technologies of cleaning of soil territories and water areas of oil pollution. *Environment Protection in an Oil and Gas Complex*, 5, 72-5.

Maksimenko, A. P. & Gersh V. A. (2006). Assessment of pollutants of the soil – sludge depots. *AgroXXI*, 7-9, 47-8.

Pikovsky, Iu. I., Veselovsky, V. A., Vshivtsev, V. S., Ernestova, L. S. & Biya, L. A. (1985). Geochemical and ecological studying of oil streams in a zone of humid subtropics. *Migration of the Polluting Substances in Soils and Adjacent Environments*. Gidrometeoizdat Publishing House, Leningrad, pp. 64-9.

Serkovskaya, G. S. (2011). Content of benz-α-pyrene in samples of commodity oils. *Chemistry and Technology of Fuels and Oils*, 3, 56.

Sverlova, L. I. & Voronina, N. V. (2001). *Pollution of Environment and Ecological Pathology of the Human*. OOP KKGS Publishing House, Khabarovsk, p. 216.

Zholdakova, Z. I. & Belyaeva, N. I. (2015). Danger of water objects pollution at oil production. *Hygiene and Sanitation*, 1(94), 28-31.

In: Ecological and Biogeochemical Cycling … ISBN: 978-1-53612-081-3
Editor: Vladimir N. Bashkin

Chapter 14

Geoecology and Geopolitics in the Arctic Region: Ecological and Political Risks and Challenges

Olga P. Trubitsina[1] and Vladimir N. Bashkin[2,3]
[1]Northern (Arctic) Federal University, Arkhangelsk, Russian Federation
[2]Institute of Physicochemical and Biological Problems of Soil Sciences of Russian Academy of Sciences, Pushchino, Russian Federation
[3]Institute of Natural Gases and Gas Technologies, Gasprom VNIIGAZ LLC, Razvilka, Russian Federation

Abstract

The article is devoted to the issues of geoecology and geopolitics in the Arctic. The authors discuss the importance of addressing both the geoenvironmental (GER) and geopolitical (GPR) risks related to industrial development of the Arctic territory. Particular attention is given to the issues of risk ratings for oil and gas companies' activities in 2009-2015, focusing on key problems, ways to turn risks into opportunities, and on the importance of interrelation and mutual influence of GERs and GPRs in order to develop a risk management strategy and a model for business planning and operations in the Arctic Region (AR).

Keywords: geoenvironmental risk, geopolitical risk, geoecology, geopolitics, rating, oil and gas industry, the Arctic

INTRODUCTION

At present, the global business community is working to achieve an efficient risk management. On the one hand, this is a well-studied matter. However, despite the variety of fundamental and applied research, it often lacks novelty and fails to address the industrial specifics of risk management and to consider the interrelations and mutual effects of various risks (Sozayeva 2016).

For oil and gas companies, risk management is also a relevant problem that has its specifics particularly in as far as their polar activities are concerned. Therefore, top managers of oil and gas companies revise their business strategies, planning and operation models for risk management (The oil and gas industry 2011) in the AR. Today, this is a subject that is being watched carefully due to the geopolitical interest that many countries have in view of the natural resource potential of the Arctic. The international community views its vast resources as something that should be shared by the entire humanity in order to meet the energy needs of the planet that is expected to reach a population of 9.1 bn by the year 2050 (Zalyvsky 2013).

By now, in addition to the development programs that are already extracting offshore hydrocarbons, such as Prirazlomnaya platform, Gazprom's affiliated company, Gazprom Neft, is preparing new programs for the upcoming years. For instance, it is planning to extract gas in the Kara Sea and Ob-Taz Bay as well as in the Pechora Sea (Safonov et al. 2008; Bashkin et al. 2015, 2016; Trubitsina & Bashkin 2016). As a longer-term perspective, the oil and gas companies also consider the Barents Sea (Bashkin et al. 2015, 2016; Trubitsina & Bashkin 2016).

The relevance of geopolitics increases significantly with consideration of the generally harsh environmental and climatic conditions in the Arctic regions rich in oil and natural gas resources. This pushes the research of

GERs for different oil and gas industry facilities (Samsonov 2007). The GER indicators are defined as risks which occur in the environmental and climatic conditions and the geopolitical conditions of the Arctic in the "industry - environment" system, as related to the mutual impacts of industrial facilities on the environment and of the environment on industrial facilities (Bashkin 2014, Trubitsina & Bashkin 2016). The evaluation of GERs reflects the nature and extent of interaction in these reciprocal relations between man-made and natural factors (Trubitsina 2016). At the same time, the global level of GPRs is associated with the worldwide processes and trends whereby international access is claimed to the Arctic zone and its natural resource potential. A disturbance of the status of strategic stability in the geostrategic domain of the Arctic is considered to be an indicator of possible GPR effects. In this context, the GPR is the probability of a change in the geopolitical situation at the regional and global levels, which can lead to adverse conditions (risk of a hybrid war, military conflicts, etc.) or additional opportunities (Ezhiev 2009).

This article deals with the oil and gas companies' key risks in 2009 – 2015. Identification of these risks by using the content analysis method allowed to rating the ten key risks in the oil and gas companies' activities, which reflect the industry's key problems and crucial development trends. The article focuses on two aspects:

1. The top rating items, which are risks associated with geoenvironment and geopolitics: a) the item "Health, Safety, and Environment and Legal Compliance Risks" (HSE) is related to the GER; b) the item "Access to Reserves or Markets: Political Limitations and Competition for Proven Reserves" (ATR) is related to the GPR;
2. The oil and gas industry's GERs and GPRs in the AR, including identification of adverse conditions and additional opportunities.

Trends of the Risk Ratings for the Oil and Gas Industry Companies

The oil and gas companies' activities are affected by adverse conditions on a global scale. As a result of this, risks arise that are long-term in nature. However, the degree of their relative significance varies throughout each year (The oil and gas industry 2011). The risk ratings are understood by the authors as the combination of all the factors which cause the risks, listed according to their item numbers which show the level of criticality and importance for the oil and gas industry companies. The major trends of the risk ratings are demonstrated in Table 1 which was prepared by the authors based on the research in business risks conducted by Ernst & Young audit and consulting company (The oil and gas industry 2011, 10 major business risks 2010, The pulse of business 2013). Ten key industry risks are ranked based on their degree of importance as of 2009, 2011, 2013 and 2015.

Access to Reserves or Markets: Political Limitations and Competition for Proven Reserves (ATR)

One of the main problems that the oil and gas industry companies have is access to sufficient hydrocarbon reserves. The risks involved in this are related to geographical, environmental, and geopolitical factors. Advanced technologies allow to developing new reserves. The growing global demand for energy and decline in production in a number of oil and gas regions that have been in operation for a long time push exploration activities in existing fields and draw attention to improvement of oil recovery. At the same time, an increasing number of companies are viewing the possibility of extracting reserves, which were previously considered unprofitable due to complicated geological or climatic conditions (The pulse of business 2013).

Table 1. Trends of the risk ratings for the oil and gas industry companies

Risk: position in the ranking	Years			
	2009	2011	2013	2015
HSE	-	5	1	1
Price volatility; managing long-term investments under the risk of significant price fluctuations	3	9	2	2
ATR	1	1	3	3
Rising costs and inflation	4	3	4	6
The uncertainty of energy policy	2	2	5	5
The deterioration of financial conditions of company's activity	5	4	6	4
Staff resources constraints	6	6	7	7
Competition caused by the emergence of new technologies and sources of energy (e.g., alternative fuels)	-	10	8	8
Information security	-	-	9	10
The increasing scale and complexity of projects	-	-	10	9
Climate change and environmental problems	7	8	-	-
Duplication of services offered by international oil and oilfield service companies	8	-	-	-
The disruption in the supply	9	-	-	-
New difficulty operating nature, including those associated with work in unknown conditions	-	7	-	-

Note that the leading items of the risk rating for 2009-2015 are the two key items: ATR and HSE. In terms of analysis of the major trends in the risk ratings, the fact to note is the change of the No.1 item in 2013 from the ATR risk to HSE. These two dominating risk items are associated with the geopolitical and geoenvironmental factors. Prior to 2011, the top risk for the oil and gas industry had been the GPR, which lost some ground by 2013, giving way to the GER. ATR is a strategic risk related to the nature of interaction with clients, competitors and investors. HSE is an operational risk, which affects the environment and processes, systems, personnel, and the company's value creation chain in general (The oil and gas industry 2011).

The need to look for new deposits and move exploration to areas that are ever more difficult to access not only increases the GPRs but also results in higher GERs. An active development of technologies to improve the efficiency of exploration and extraction of new reserves opens new horizons for the oil and gas industry and essentially disproves the widespread assumption that the peak of production is behind us.

Exploration and development of fields in new regions is for many companies the only chance to increase the level of reserve replacement. In the conditions of a permanent search for new sources of oil and natural gas, the plans to explore and extract offshore hydrocarbon reserves in the Arctic begin to have a more realistic outlook (The oil and gas industry 2011; The pulse of business 2013). For international oil companies, the profitability of their activities here depends on being able to gain stable access to hydrocarbon reserves (10 major business risks 2010) and thereby minimizes the GPRs.

Health, Safety, and Environment and Legal Compliance Risks (HSE)

This risk first appeared in 2011 and went straight to the fifth position in the top ten. According to the respondents' opinions at the time, the impact of the HSE risk on the oil and gas industry would not have changed in a significant way over the next three years (The oil and gas industry 2011). However, in 2013 a shift occurred in the risk distribution pattern, and the HSE item hit the top spot in terms of importance.

The HSE issues became more relevant to the oil and gas companies as their top managers began to pay closer attention to a sustainable and resilient model of activity, also reflecting more complex operational challenges facing the industry. This implicates a higher influence exerted by the public (10 major business risks 2010). For instance, in the approaches used to assess the external component of investment attractiveness of enterprises, increasing the value of environmental ratings as the element of development business responsibility for the implementation of the "trinity" principle of economic, social and environmental aspects of the company. Therefore, the introduction of environmental ratings has made it possible to prioritize attracting of investments to those companies, which the most meet the sustainable development criteria. The growing interest of investors (both institutional and private) towards ratings provokes companies to view them as a drive to continuously improve their internal environment (Trubitsina & Bashkin 2016, Trubitsina 2014).

Close attention to the HSE issues can affect the oil and gas companies' ability to explore and produce in the AR, which is a highly environmentally sensitive area. For example, at present there is a growing concern among the public and political community about possible environmental consequences of oil and gas field development there and, therefore, about higher GERs. Thus, the oil and gas companies are under thorough scrutiny not only by the government, but also by the shareholders and the public political community, which demands disclosure of the GER information. In the aftermath of the environmental disaster in the Gulf of Mexico involving a major oil spill, complete disclosure is now expected of all information related to offshore drilling and possible environmental impacts of drilling activities offshore and the steps taken by the oil and natural gas companies to prevent similar accidents, minimize and control the GERs. The global community's concern about the condition of the environment affects the oil and gas companies' management decision-making process with regard to strategic development, currently putting to the forefront objectives such as mitigation of the GERs by improving the risk management strategy, the companies' investment appeal, etc. (The oil and gas industry 2011).

The Transformation of the GER and GPR in Additional Opportunities

Oil and gas industry risks can be turned into opportunities by a number of ways, of which the key ones include: 1. Exploration and production in new regions. Expansion of the geographic boundaries of operations opens new opportunities for companies which have the necessary technologies and resources. 2. Conventional field development in hard-to-reach regions. The respondents divide such regions in two groups. One includes countries such as Iraq or Nigeria where investors have to deal with significant political risks and security risks, the other group includes the Arctic and other environmentally vulnerable regions where operators face the challenge of having to reduce the environmental impact (The oil and gas industry 2011).

These ways of transforming the GERs and GPRs into opportunities can be very advantageous if the interaction between geopolitics and geoenvironment is enhanced. This interaction can produce a powerful effect if emphasis is put on the concept of sustainable development (Gosteva 2010). The oil and gas companies are currently developing the AR which was previously regarded as too hard to reach, too costly or politically unstable. With this in mind, operating here did not appear to be a feasible option. However, given the anticipated high demand for energy and the availability of advanced technologies and business processes, the region starts sparking interest. Indeed, exploration and production in the AR provide key opportunities for oil and gas companies, and development of technology appears to be the main reason behind this trend.

It is believed that the importance of these opportunities will only increase. The availability of new technologies, improvements in the quality of seismic data and more efficient interpretation methods have recently made it possible for oil and gas companies to access a region as undeveloped as the Arctic. However, while innovative technologies designed to reduce the GERs provide access to additional reserves, companies within the industry are fighting for licenses to explore and produce in new regions rich in mineral resources, as well as for the technologies themselves, which increases the GPRs. Many companies have already signed agreements for joint operations and other forms of cooperation in order to access new technologies and discover opportunities for transferring them to other markets. In this regard, companies should be more flexible in selecting their business strategy, and should invest in the development of both cooperation and the new region (The oil and gas industry 2011). Thus, exploration and development of conventional fields in regions that are hard to reach due to political, natural, environmental or other reasons can be opportunities for business development. The degree of importance of these opportunities can be viewed differently. On the one hand, investors face serious GPRs and security threats with a high degree of uncertainty. On the other hand, they have to minimize the GERs in the environmentally sensitive AR.

In order to use these opportunities to the full extent, companies have to invest in development of technologies and human resources and come up with adaptable business models. The advancement of technologies both increases the oil recovery rate and reduces the GERs.

Disregard for the GERs can have the direst consequences in the form of huge penalties and an irreversible loss of reputation in the eyes of the public. The last ten years were marked by tightening legal and regulatory requirements in response to a growing level of the GERs related to HSE (The pulse of business 2013). Compliance with the requirements in a situation where the laws are changing all the time can be challenging, especially given the fact that all countries have their national requirements and their own way of interpreting them. However, there is a growing understanding across the globe that regulatory requirements need to be harmonized and a uniform international mechanism should be established to regulate businesses. This process has its drawbacks. One of them is a growing risk related to the uncertainty in interpretation of uniform international requirements and the mechanisms for their application.

To address this problem in relation to the oil and gas companies' activities in the Arctic, Gazprom VNIIGAZ Ltd held the first meeting of the newly formed Subcommittee WG8 "Arctic Operations" of the Technical Committee ISO/TC "Materials and Equipment for Petroleum, Petrochemical and Natural Gas Industries" of the International Organization for Standardization (ISO) in Moscow (November 2012). The meeting reviewed suggestions from polar countries regarding the development of standards under the auspices of the ISO. It was also attended by offshore production countries: Kazakhstan, Italy, France, etc. The meeting approved the suggestion to develop a standard for "Environmental Monitoring in Offshore Production" to provide a standardized system for monitoring of the environment condition and measurements made as part of oil and natural gas activities in the Arctic region. Measurement of certain key indicators which characterize the extent of impact and the condition of environment components near oil and gas production facilities forms the informational basis for evaluating and predicting the development of ecological situation in order to make management decisions needed to minimize the GERs (Bukhgalter & Tikhomirov 2013). A Working Group (WG3) of experts from countries including Russian Federation (RF), Italy, Canada, Netherlands, Norway, France was established. The author, O.P. Trubitsina, also became a member of the WG3 group on behalf of the RF. It was expected that the experts' joint work on the document would allow to considering the international and national specifics of environmental monitoring in the Arctic region of offshore production, and assist in minimizing the GERs in general. However, the GPRs associated with the sanctions against the RF have made an impact on the expected uniform international document for minimization of GERs in the Arctic that would take into account the interests of all concerned countries. In 2015, the Russian experts received a letter from the International Association of Oil & Gas Producers (IOGP) which stated that «...Imposition of international sanctions, against the Russian Federation, has imposed legal restrictions on how IOGP engages with the International Organization for Standardization ("ISO") process...This process excludes participation of experts from embargoed countries...Accordingly, in order that IOGP complies with its legal obligations set out in the EU Sanctions, and safeguards the IOGP's members in respect of complying with their legal obligations set out in both EU and US law..., IOGP has decided that it is legally obliged to terminate the engagement of experts from Russian companies and Russian Institutions with its Standards» (Engagement of the International Association 2015). Thus, in this situation the GPRs were more significant and affected the mechanism for establishing a common international standard to minimize the GERs in the AR for the good of all concerned parties, including Russia.

The Arctic as a Territory of Strategic Interests: Threats to the GER and GPR

The Arctic is an area where strategic interests of a number of countries cross each other. This region holds an exceptionally important military strategic position and has international significance in terms of economics, resources, transportation, and logistics (Kolesnik 2015). Opinions on the perspectives and risks of the Arctic region in global terms have never been as polarized as they are now, ranging from claims that "the Arctic is the last confine of pristine nature which requires protection from the pressures of modern civilization" to "a new energy outpost that will provide energy security, prosperity and jobs on the Arctic countries' shoreline." Thus, the Arctic is at the same time a region of immense perspectives and great paradoxes. One of the main dilemmas is that, on the one hand, the Polar region (as no other place in the world) is under a serious threat of climatic changes, as the scientific research indicates. On the other hand, it is the climate change and the melting of ice as a result of it that opens the Arctic to the big industry (Resource development 2012).

This presents a wide range of threats and challenges for the national interests and security of the RF. It includes other countries maintaining a stable interest in the Arctic zone of the Russian Federation (AZRF) and the Northern Sea Route (NSR), which is reflected in the activity of governmental and non-governmental organizations, mainly of the US and Norway, with support from the political and business communities of the West, as well as Japan and China, pursuing projects intended to increase influence on the border regions of the RF, with the goal of transforming the NSR into an international transit line (Kolesnik 2015).

The discussion of the NSR status is a problem issue for Russia's dialog with the world. It is probably fueled by the assumption that the "leading global masters are drafting their own scenario for the control of the circumpolar space for Russia. A part of it is the West European countries' efforts to solve the problem of dependence on Russian energy" (Chumakov 2009). This information environment is being created to "erode the stability of Russia's Arctic behavior" as, for instance, nobody is trying to contest Canada's regulation of shipping in the Northwest Passage (within the Canadian sector) (Zalyvsky 2013). The future issues involve the potential resources of the Arctic, where energy resource reserves are expected and where climate change could change the current logic of trading routes. The success of developing the resources at Sakhalin has also brought development of natural gas resources in the cold and fragile environments within closer range, but can also change the dynamic of areas of strategic interest (Geopolitics and Natural Gas 2012).

Thus, the AR is a territory of strategic interests, "…with the discovery of petroleum and natural gas deposits in the Arctic, successful exploration for solid mineral resources and, hence, various territorial claims by the Arctic nations…" (Kovalev 2003). Delimitation of borders by states is a necessity caused by development of the humanity (Menon 1979). Disputes associated with delimitation of sea areas have been on the rise lately. This is indicated, for instance, by the growing number of claims filed with the International Court of Justice (Malanczuk 1997). The number of disputes will continue to increase (Irwin 1980), due in large part to the growing interest that all countries exhibit in oil and natural gas reserves and other natural resources located in the world's oceans, as well as the climate change (Radysh 2006). The International Court of Justice notes that "the justice does not attempt to make equal what the nature made unequal" (ICJ 1985, Trubitsina & Gorbunov 2016).

Based on the materials (Smirnov 2016), the authors have prepared Table 2, which reflects a rather conditional classification of countries according to their geostrategic relation to the Arctic. The three groups that were identified compete both between each other and as part of international organizations.

In the Arctic regions, oil and gas exploration has increasingly gain attention by some of the main geo-strategic players (i.e., US and Russia), due to climate change and melting ice in the region. For the next few decades, Arctic oil and gas exploitation will be predominantly in the Exclusive Economic Zone (EEZ) of the first group's countries and perhaps further into the Arctic Ocean outside or close to the Arctic Circle. This involves a large amount of proven and estimated oil and gas reserves, primarily to be found within the Russian borders. There are still uncertainties concerning the legal status of the Arctic region, which will undoubtedly increase the current attention of the main geo-strategic and regional players for geo-economic and geopolitical purposes (Geopolitics and Natural Gas 2012).

Table 2. Groups of countries according to their geostrategic relation to the Arctic

Group number			
The first group		The second group	The third group
Countries	The Arctic five: Russia, USA, Denmark, Canada, Norway (the latter four being NATO members)	Subarctic countries which do not have access to the Arctic Ocean but are members of the Arctic Council: Iceland, Finland, and Sweden.	Non-Arctic countries: Brazil, India, China, Singapore, South Korea, Japan, EU countries, etc.
Features	All of these have the right to develop the offshore natural wealth, which has become accessible by virtue of global warming and the use of innovative technologies. These countries have unsettled conflicts. For instance, this includes northward expansion of the Arctic shelf.	These do not have the rights to the shelf area but try to promote their status and influence through the Arctic Council. Iceland is a NATO member, while Finland and Sweden have ramped up the discourse for joining the NATO, which was confirmed by these countries' leaders attending the NATO summit in Warsaw in June 2016.	Some of these obtained the permanent observer status in the Arctic Council in 2012. Their objective is to maximize their status and try to influence a revision of the Arctic's status by including it in the global commons.

The analysis of the Arctic countries' goals and actions shows that they are aimed at proving that Russia has no legal backing for a claim to develop offshore natural resources and to use the NSR as an internal passage, and blame Russia as incapable of ensuring environmental safety in production of natural resources in the region (Smirnov 2016). These aspects show the hybrid nature of the threats that Russia faces in the Arctic.

At present, there is no common understanding of the term "hybrid wars" which denotes a coordinated use of political, diplomatic, informational, psychological, economic tools and force to achieve strategic objectives. However, NATO experts already use the notion of "hybrid wars" when referring to Russia's role in crisis areas (Klimenko 2015).

Russia's activity in the AR strikes a nerve with the Western partners and provides an occasion for individual representatives of the political circles, the military establishment and security agencies to speculate on the Russian threat. For instance, the US Director of National Intelligence, J. Clapper, in the annual report to Congress on global security threats February 9, 2016, called Russia and its increasing presence in the Arctic one of serious threats in the region, especially if relations with the West continue to deteriorate (Gordeyev 2016).

Let us review some of the aspects that the hybrid nature of threats facing Russia in the Arctic is manifested through.

1. The "double standards" applied by the NATO and the US to interpreting some international treaties (including UN treaties) and ignoring others related to the Arctic. For instance, in 2009, long before the Syria and Ukraine crises, the double directive for outside and inside US national security "Policy in the Arctic Region" signed by the US President stated that "the US has broad fundamental interests in the Arctic for national security and is ready to act accordingly alone or in collaboration with other countries to protect these interests" (Smirnov 2016). A priority of the "National Strategy in the Arctic Region" approved in the US in 2013 is enhancing national security in the region and creating a defense infrastructure. The "US Navy Arctic Roadmap 2014-2030" developed in 2014 reflects the possibility of military conflicts in the Arctic and specific objectives for the Navy to ensure American dominance. In the National Military Strategy published by the US Department of Defense in 2015, Russia is referred to as a revisionist state that challenges the international norms, does not respect the sovereignty of its neighbours, and undermines regional security (Smirnov 2016, Gordeyev 2016).

The US security agencies zoomed in on the Russian Arctic in 2014, by establishing a strategic council for expert information exchange. Data from the US intelligence satellites and deep-water transmitters is used for problem monitoring in the aerospace, marine, inland, and cyber space in the Arctic. In 2017, the US National Geospatial Intelligence Agency is planning to publish 3D maps of the entire Arctic, which were collected under the good pretense of environmental monitoring (Gordeyev 2016). In this case, the GER is a subject of geopolitical manipulations. The US leadership uses such documents to develop specific plans for expansion of marine operations to ensure the US dominance in the Arctic Ocean. It is known that the sanctions against Russia, declared due to the Ukraine crisis, began with a number of resolutions of the US government and their satellites, establishing criminal and financial penalties for top managers of companies which supply equipment and technologies for hydrocarbon production in the AR to Russia. This is an attempt to undermine the process of Arctic development by Russia (Smirnov 2016). By 2020, the US troops are expected to be fully prepared for action in the Arctic latitudes (Gordeyev 2016).

2. Geopolitical provocations against inability to ensure environmental safety in the Arctic are also associated with manipulation of GERs in geopolitical interests. A number of such provocations have been made, and they are unprecedented. An example of that are the widely advertised statements by the WWF that oil production and transportation in the Arctic present an unacceptable risk until Russia puts in effect new laws that will "guarantee sufficient safety of oil operations in the sea." The usual concerns about increasing load on the Arctic ecosystem are voiced regularly by the notoriously known international environmental organization "Bellona" (Smirnov 2016). Greenpeace activists also cast doubt on companies' experience of drilling in harsh climate conditions, citing slogans such as "Protect the Arctic," "Arctic is not a place for oil production, industrial fishing, and wars," etc., exemplified by the situations in 2012, 2013 involving action against oil production in the AR on Prirazlomnaya platform. This is an ultimatum declared to Russia, requiring it to stop any commercial activities in the Arctic.

Conclusion

Today, the main risks for the oil and gas industry are risks related to geoenvironment and geopolitics. These are: HSE and ATR. These risks can be transformed into additional opportunities by a number of ways. For example, this can be done through exploration and development of conventional deposits in a hard-to-reach region such as the Arctic. In this case, expansion of the geographic boundaries of operations opens new opportunities for companies, which have the necessary technologies and resources. However, such development of the Arctic territory involves significant GERs and GPRs. At the same time, these risks can turn to negative conditions rather than opportunities. This includes the risk of a hybrid war, possible military clashes, etc. Companies should therefore be more flexible in selecting the business strategy and factor in the interrelation and interdependency of the GERs and GPRs.

From the global community's standpoint, enhancement of the interaction between geopolitics and geoenvironment appears to be very promising for transformation of GERs and GPRs to additional opportunities. This interaction can have a powerful effect if the concept of sustainable development is emphasized. As a tool for this concept, the authors propose the use of the "hawk geopolitics strategy" (Grigorieva et al. 2015) for the AR, aimed to achieve total geoenvironmental and geopolitical control of the territory as a strategic resource of the world and individual countries.

Conflict of Interest

The authors confirm that this article content has no conflicts of interest.

Acknowledgments

The authors express their deep appreciation and gratitude to their affiliation institutions for the support during research and development.

References

10 major business risks for companies in the oil and gas industry (2010) Research by Ernst & Young. Available at: www.rcit.su/article021.html.

Bashkin, VN (2014) Biogeochemistry of polar ecosystems in the gas industry impacted zones. Gazprom VNIIGAZ Publishing House, Moscow.

Bashkin, VN, Trubitsina, OP & Priputina, IV (2015) Evaluation of geo-environmental risks in zones of influence of oil and gas industry in the Russian Arctic. *Arctic and North,* 19, 92-98.

Bashkin, VN, Trubitsina, OP & Priputina, IV (2016) Evaluation of geo-environmental risks in the impacted zones of oil and gas industry in the Russian Arctic. In: Bashkin, VN (Ed) *Biogeochemical Technologies for Managing Environmental Pollution in Polar Ecosystems*, Switzerland: Spinger, 109-116.

Bukhgalter, EB & Tikhomirov, DV (2013). Development of the international standard "Environmental monitoring in offshore oil and gas production in the Arctic." *Environmental Protection in Oil and Gas Complex,* 7, 47-48.

Chumakov, DS (2009) Energy resources in the Russian Arctic in the global economy. Author. dis. cand. economy. *Sciences*. Moscow. 25 p.

Ezhiev I. (2009) Geopolitical risk as a political category. *Power,* 12, 143-146.

Gordeyev V. (2016) US National Intelligence named seven threats from Russia. Available at: http://www.rbc.ru/politics/09/02/2016/56ba0c819a794764fb74898a.

Gosteva, SR (2010) The Science of government (some issues of the geopolitics of modern Russia). *Vestnik TSU,* 2 (82), 203-212.

Grigorieva, MA, Markelov, DA, Markelov, AV, Mineeva, NY, Polynov, OE & Akol'zin, AP (2015) "The strategy of geopolitics Kite": total control over the territory as a tool for sustainable development. *Herald of the Buryat State University,* 4, 55-60.

ICJ, Reports: Case concerning the continental shelf – Malta/Libya. The Hague, 1985.

Klimenko, S. (2015) Theory and practice of "hybrid warfare" (NATO views). *Foreign Military Review,* 5, 109-112.

Kolesnik, AA (2015) Features of the modern concept of border activities in the Arctic region. *Proceedings of annual scientific-practical conference of Northern enterprise Institute. Arkhangelsk: KIRA.* 88-96.

Kovalev, AA (2003). The modern international maritime law and practice. Scientific book. Moscow.

Menon, PK (1979) Settlement of international boundary disputes. *The Anglo-American Law Review,* 8 (1), 17-23.

Radish, MB (2006) Legal problems of the delimitation of Maritime spaces in the Arctic. *Bulletin of PFUR. Series "Juridical science,"* 1(9), 170-176.

Resource development in the Arctic: risks and responsible management. Report of the Fridtjof Nansen Institute and DNV, presented during the summit, ONS 2012 in Stavanger, Norway, 2012. Available at: http://pro-arctic.ru/category/resources/page/7.

Safonov, VS, Volkov, AN, Kovalev, SA & Lesnych, VV (2008) Categorization of objects of JSC "Gazprom" according to the degree of potential hazard: theory and practice. In: Safonov, VS (Ed). *Industrial and Environmental Safety of Gas Industry.* Gazprom VNIIGAZ Publishing House, Moscow, 151-163.

Samsonov, RO (2007) Assessment and management of geo-ecological risks in the gas industry. *Oil and Gas Business,* 1. Available at: http://ogbus.ru/years/12007/.

Sozayeva, DA (2016) Risk Management: approaches, models, methodology. *Issues of Risk Analysis.* 13, 6-10.

The oil and gas industry — 10 major business risks. 2011. *Research by Ernst&Young*. Available at: www.ey.com/Publication/vwLUAssets/Turn-risk-into-results-OG-RU/$FILE/Turn-risk-into-results-OG-RU.pdf.

The pulse of business life Comparative analysis of 10 key risks and opportunities in 2013 and beyond. The oil and gas industry. Research by Ernst&Young. Available at: http://changellenge.com/wp-content/uploads/2014/04/EY-Business-pulse-Oil-and-Gas-RUS.pdf.

Trubitsina, OP & Bashkin, VN (2016) The analysis of geoecological risks and ratings as a factor increasing the investment attractiveness of enterprises. *Issues of Risk Analysis,* 13. 62-68.

Trubitsina, OP & Bashkin, VN (2016) The analysis of geoecological risks and ratings as a factor of improving investment attractiveness of enterprises. In: Bashkin, VN (Ed) *Biogeochemical technologies for managing environmental pollution in polar ecosystems*, Spinger, Switzerland, 141-150.

Trubitsina, OP & Gorbunov, SN (2016) Environmental justice and fairness. In: Abashidze, AH (Ed) *Proceedings of the XIV International Congress "Blischenko readings."* People's Friendship University, Moscow (in press).

Trubitsina, OP. (2014) Investment attractiveness of environmentally oriented enterprises. *Actual problems of modern business. The materials of international correspondence scientific-practical conference*, Arkhangelsk, 82-85.

Zalyvsky, NP (2013) Geopolitics of Russia's national interests in the Arctic in the conditions of recession of economy of the Arctic countries. *Vestnik of Northern (Arctic) Federal University. Series: Humanities and social Sciences,* 3, 88-96.

In: Ecological and Biogeochemical Cycling … ISBN: 978-1-53612-081-3
Editor: Vladimir N. Bashkin

Chapter 15

ASSESSING CRITICAL LOADS AND ECOLOGICAL RISKS FOR POLAR MARINE ECOSYSTEMS

***Irina V. Priputina*[1] *and Vladimir N. Bashkin*[1,2]**
[1]Institute of Physicochemical and Biological Problems of Soil Science RAS, Pushchino, Russia
[2]Gazprom VNIIGAZ LLC, Razvilka, Russia

ABSTRACT

In this paper the main attention is paid to the proposed methodology of assessing polar marine ecosystem risks associated with industrial emissions. This approach allows to making both quantitative estimation of potential changes in the ecosystem condition and calculating the probability of their occurrence. It is also providing a detailed characterization of the Barents Sea ecosystems as targets of man-induced impact.

Keywords: marine ecosystems, critical loads, nitrogen, ecological risk exceedances, risk assessment and calculation

INTRODUCTION

The Shtokman gas field is located 550 km from Murmansk (North-East). The nearest land is the West coast of New Land (300 km). The study of the Shtokman field began in 1981. The depth of the first exploratory well was 3153 m, after the drilling two deposits were discovered. The sea depth in the area is from 320 to 340 m. The reserves of the Shtokman field are about 3.94 trillion m^3. Condensate is present in the amount of 56.1 million tons. In the bottom deposition complexes of modern soft and soft granulometric composition grounds were identified. Their capacity is up to 8 m. In addition, there are common underlying soft depositions, whose thickness is 4-24 m. The high sensitivity and vulnerability of the nature in the Russian Arctic to the anthropogenic influence are known.

In these circumstances compliance with Russian and international environmental standards, respecting the environment and preventing negative impacts on marine and terrestrial ecosystems of the Arctic zone during the construction and operation of the Shtokman gas condensate field is one of the basic conditions of the project. Furthermore, the possible changes in the biological nitrogen cycle resulting from the potential development of this gas deposit may be associated with emissions of nitrogen oxides while conducting geological exploration, extraction and transport of gas. The permissible impact of these emissions can be assessed on the basis of the methodology of critical loads and ecological risk assessment for marine ecosystems (Bashkin et al., 2011, Bashkin 2014, Bashkin & Priputina 2015).

Ecological Risk Assessment

Methods of Evaluation of the Ecosystem Risk Based on the Methodology of Critical Loads

The assessment of ecosystem risks based on critical loads of pollutants are encouraged to follow the formal procedure of risk assessment (Bashkin & Priputina 2015). At the hazard identification stage should set the emission sources, to determine possible scenarios of anthropogenic impact and to compile a complete list of pollutants contained in the emissions of the projected enterprise. In addition, it is necessary to outline the range of potential recipients of impacts (marine ecosystems within the zone of influence of the projected object) and spend their ranking. On the basis of available information on the hazards and the recipient a qualitative characterization of impacts must be defined as well as the list of pollutants for which it is advisable to carry out a detailed risk assessment (priority pollutants) should be made. Exposure assessment needs to include a detailed specification of the recipient (including ecosystems-recipients unification to receptor sites) with the establishment of a background and calculate the projected load level for priority pollutants, the magnitude of deposition of pollutants (g/ha per year, or Eq/ha per year). At the stage of assessing the ecosystem effects one should conduct a mapping and calculation of critical loads of priority pollutants, characterizing maximum permissible load level to selected recipients. Characteristics of ecosystem risks should include the calculation of changes in the condition of the recipients, the probability of their manifestation and determination of the acceptability of such changes in accordance with the selected criteria.

It is proposed to conduct the risk characterization in two phases. The first step is to carry out a deterministic calculation of critical load exceedances for pollutant X [Ex(X)] based on the averaging the input data. In case of detection of receptor plots Ex(X)>0 in the second stage, it is advisable to assess ecosystem risks using modeling methods (see Bashkin & Priputina 2015).

Studies to assess ecosystem risks should be completed by uncertainty analysis of the obtained results. To do this you must describe the sources of uncertainty at each stage of the risk assessment and evaluate the reliability of calculation results. The results of the assessment of ecosystem risk is proposed to use to rank the individual design alternatives and to develop approaches for mitigation of environmental impacts in the evaluation procedure of impact on environment of planned economic activities.

Therefore, the proposed risk assessment method for ecosystems related to emissions of the enterprises of the gas industry, allows us to quantify not only the magnitude of the projected changes in marine ecosystem conditions, but also of their probability. It allows for detailed features of ecosystems as objects of technogenic impact. In addition, this method takes into account the close relationship between the individual components of terrestrial and aquatic ecosystems as well as natural variability of parameters characterizing the state of these components.

Thus, the quantitative estimates of ecosystem risk are advisable in the preparing the environmental impact assessment (EIA) study of projects in the gas industry, located in areas of low information data and a high degree of uncertainty.

Marine Ecosystem Risk Assessing for the Shtockman Project Impacted Area

Ecological risk assessment for the zone of influence of NO_x emissions during operation of the Shtokman gas field is served to the maintaining optimum ecological and trophic conditions for phytoplankton, which can be considered as the initial link the biological nitrogen cycle in marine ecosystems. The risk profile changes in the trophic status of marine waters in the area of development of phyto - and zooplankton are estimated in accordance with the following scenario. This seems reasonable that in winter (8 months a year) nitrogen deposition are deposited in the surface snow and ice cover, whose thickness could reach 2-5 m. The calculated concentrations of nitrogen inputs to the waters (levels of exposure), corresponding to the model parameters of deposition of nitrogen is presented in Table1. Nitrogen accumulated on the surface of the waters during the winter period determines short-term "peak" eutrophication of marine ecosystems, because the melting of the ice coincides with the spring mass development of phytoplankton (Zenkevich 1963). Because at that time the surface marine water is not limited by phosphorus, the nitrogen may be included in the biological cycle in the composition of phytoplankton, and saturation of surface waters with oxygen (100%) is a "guarantee" of preservation in the favorable oxygen regime.

Table 1. Nitrogen content increase in the sea ice-snow cover as a result of increasing levels of nitrogen deposition during operation of the Shtokman field

N loads (kg N/ha/yr.	N content increase in sea water layer (mkg/l/day)		Total N content increase in sea water layer for summer period (mkg/l)		N accumulation in ice-snow cover for winter period (mg/m^3)	
	0-10 m	0-50 m	0-10 m	0-50 m	0-2 m layer	0-5 m layer
5	0.14	0.03	16	3	168	67
7.5	0.21	0.04	25	5	252	101
10	0.27	0.05	33	7	336	134
15	0.41	0.08	49	10	503	201
20	0.55	0.11	66	13	671	268
30	0.82	0.16	99	20	1007	403
50	1.37	0.27	164	33	1678	671
100	2.74	0.55	329	66	3356	1342
150	4.11	0.82	493	99	5034	2014
200	5.48	1.10	658	132	6712	2685
400	10.96	2.19	1315	263	13425	5370

In the summer (4 months a year) due to nitrogen inputs from atmospheric deposition to the surface waters and mixing of the upper layer of water there is a gradual increase in nitrogen concentrations in the upper part of the sea depth (the mixing depth 10-50 m). The calculations show that total during the growing season concentrations of nitrogen (excluding consumption by biota) depending on the values of deposition loads and the water column layer can be from 3-16 to 260-1300 μg N/l (Table 1). According to field observations (Zenkevich 1963), during this period, unpolluted surface waters have almost no nitrates. Thus, the calculated nitrogen concentrations (represented mainly by nitrate forms) will determine the trophic state of the waters in the area of development of plankton communities.

Based on the selected indicator criteria and the Table 1 data at the level of deposition equal to 10 kg N/ha per year the environmental risks of eutrophication for communities of plankton in this area of the Barents sea can be considered negligibly small. The level of deposition of 10-50 kg N/ha per year can be considered as low risk. When applying more than 50 kg N/ha per year the trophic status of waters can change and the appearance/increase of the proportion of hypereutrophic species in the biotic community while reducing oligotrophic species is possible. According to the analysis of the presented data and the simulation results of the levels of deposition of nitrogen at the planned volume of gas production (Figure 1), with the highest probability of ecological risk of eutrophication zone plankton can be observed within a 10-kilometer zone from the source of emission (worst case scenario) as a result of NOx emissions during the development of the Shtokman field.

However, the calculations do not take into account the movement of water masses (and natural in this case, dilution of nitrate concentrations), and active fixation of available nitrogen in the phytoplankton biomass, lowering its content in marine water. According to the available expert estimates for the North sea, the total daily intake of nitrogen deposition at the level of 0.2 μg N/l per day could increase phytoplankton biomass by an amount corresponding to 1.2 mg C/m^3 (Hessen et al. 1992). Considering that all of the nitrogen received from atmospheric deposition during the summer period, can be absorbed by biota, it is possible to calculate the increase in supplies as the mass of phytoplankton at different levels of deposition of nitrogen (Table 2).

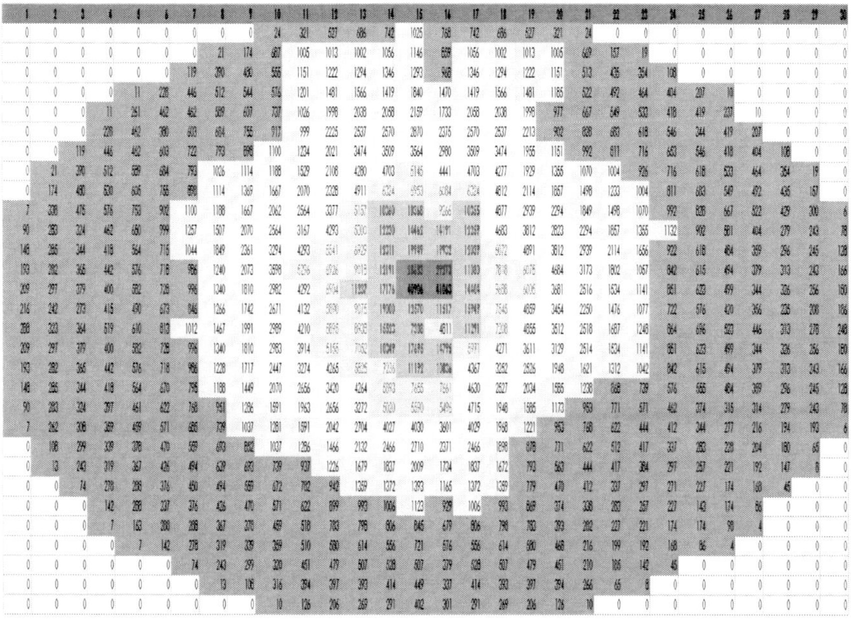

Figure 1. Levels of anthropogenic deposition of nitrogen to the waters of the Barents sea.

Taking into account the elemental composition of phytoplankton, in which the share of nitrogen accounts for about 10% (Zenkevich 1963), stocks of fixed N in phytoplankton biomass in terms of an average area over the summer period will increase to the zone of maximum impact at 1-1.5 kg N/ha. In the presented calculations, increase productivity and stocks of phytoplankton as a result of increased atmospheric supply of nitrogen is not considered the typical for the summer period P limiting for involvement of nitrogen in the biological cycle. Sources of phosphorus for phytoplankton are rising to the surface waters in autumn and winter, owing to deep water upwelling and coming from atmospheric deposition. The literature presents conflicting data regarding the atmospheric delivery of phosphorus. Thus, in some publications the ratio N/P in precipitation of the background territories is estimated at 150:1 that the level of nitrogen deposition to 5 kg N/ha per year (background flow in the Arctic sea or 500 kg N/ km^2) corresponds to only 3.5 kg P/ km^2 per year. For marine areas of Northern Europe a value of 10 kg P/km^2 per year was indicated (Hessen et al. 1992), taking into account the contribution of marine aerosols to the total component of the fallout and this seems more reliable. Using "normal" phytoplankton atomic ratio C: N: P = 106: 7: 1 (Redfield et al. 1963), one can calculate the amount of nitrogen that can be assimilated by phytoplankton during the summer season due to the atmospheric delivery of phosphorus. It is less than 1 kg N/ha, which determines almost the entire zone of influence of the Shtokman field the risk of imbalance of N/P for phytoplankton at the end of the growing season, when it will be spent "deep" supply of phosphorus into the surface layer of marine waters.

Table 2. The increase in biomass of phytoplankton as a result of increasing levels of nitrogen deposition during operation of the Shtokman field

Loads (kg N/ha/yr	Additional N input for 0-50 m layer (mkg/l)	*C stock increase in phytoplankton biomass for 0-50 m layer (mg C/m³ d.w.)	Additional N input for 0-10 m layer (mkg/l)	C stock increase in phytoplankton biomass for 0-10 m layer (mg C/m³ d.w.)
5.0	3	0.4	16	2
7.5	5	0.6	25	3
10	7	0.8	33	4
15	10	1.2	49	6
20	13	1.6	66	8
30	20	2.4	99	12
50	33	4	164	20
100	66	8	329	39
150	99	12	493	59
200	132	16	658	79
400	263	32	1315	158

* In the calculations was used the characteristic for the Barents sea ratio between reserves and productivity of the phytoplankton biomass equal to 50 (Zenkevich 1963).

RISKS OF DISTURBING ENVIRONMENTAL AND TROPHIC STATUS OF WATERS

Risks of disturbing environmental and trophic status of waters in the area of development of benthos in relation to impacts associated with increasing nitrogen inputs are determined primarily by the effects of "oxygen deficit", which is determined by the increase in the consumption of free oxygen for oxidation of organic matter originating by deposition of dead biomass of plankton, and also with the oxygen "expenditure" on denitrification. According to available data, for oxidation of 1 kg of zooplankton (wet weight) it is necessary 222 g of O_2. According to available data (Zenkevich 1963), zooplankton in the open part of the Barents sea is 200-2000 mg/m^3, which for a 50-meter layer is 40-400 kg/ha. Possible increase in zooplankton pool owing to the increasing phytoplankton production at current levels of deposition of nitrogen, according to our estimates, can reach 10%, increasing the magnitude of the flow rate of oxygen required to oxidize organic matter after the death of the biota.

Denitrification in marine ecosystems is estimated as 0.1-4 mg N/m^2 / day (or 0.3-0.4 to 14-15 kg N/ha per year) and accordingly it "requires" for creating 1 mol of N_2 up to 138 g of oxygen. Being compared to the background levels multiple increase of nitrogen deposition in the area with the greatest impact on the Shtokman field will lead to higher concentrations of dissolved nitrogen in marine waters and will contribute to intensification of denitrification, thus increase the risk to ecological status of benthic community. However, as in the case of phytoplankton, the spatial area of potential risk to benthic communities in the first years of potential exploitation of the Shtokman field will be limited to a water area with a radius of 10-15 km (20-25 km taking into account the influence of ocean currents). In the long run, to benthic communities are not excluded environmental violations determined by the cumulative effect of the increased supply of nitrogen for several years.

Conclusion

Performed assessments of potential risks to marine ecosystems are qualitative in nature, identifying levels of stress, which may cause environmental hazards violations of the trophic structure of the individual sections of the Barents sea. However, in the absence of other available data, they provide the minimum necessary information for making management decisions to reduce potential environmental risks in connection with the development of the fields in shelf zone of the Arctic seas.

Conflict of Interest

The authors confirm that this article content has no conflicts of interest.

Acknowledgments

The authors express their deep appreciation and gratitude to Dr A. Tankanag (Institute of Cell Biophysics RAS) for the consultation and active participation during result discussions.

References

Bashkin, V. (2014) Biogeochemistry of polar ecosystems in gas industry impacted zones. *Gazprom-VNIIGAZ*, Moscow.

Bashkin, V. N., Priputina, I. V. & Tankanag, A. V. (2011) Estimation of geoecological risks in the production and transportation of LNG, *Science and technics in the gas industry*, 1, 107-112.

Bashkin, V. N. & Priputina, I. V. (2015) (Geo) Ecological Risk Assessment in Gas Industry Development Scenarios, *The Open Ecology Journal,* 8, 65-68, Electronic publication date 14/9/2015 [DOI: 10.2174/18742 13001508010065].

Hessen, D., Vadstein, O. & Magnusson, J. (1992) Nitrogen to marine areas, on the application of a critical load concept. In: Thornelof, E (Ed), Critical Loads for Nitrogen. Nord 41, Nordic Council of Ministers, 161-237.

Redfield, A. C., Ketchum, B. H. & Richards, F. A. (1963). The influence of organisms on the composition of seawater. In: Hill, NM (Ed) The sea, vol 2, Wiley, New York, 26-77.

Zenkevich, L. A. (1963) Biology of USSR seas, USSR Academy of Science Publishing House, Moscow.

In: Ecological and Biogeochemical Cycling … ISBN: 978-1-53612-081-3
Editor: Vladimir N. Bashkin

Chapter 16

ASSESSING CRITICAL LOADS FOR NITROGEN UNDER CLIMATE CHANGE BASED ON CHEMICAL AND BIOLOGICAL INDICATORS IN A SUB-ARCTIC COUNTRY: THE CASE OF SWEDEN

***H. Sverdrup*[1,2], *S. Belyazid*[3], *C. Akselsson*[4] *and M. Posch*[5]**
[1]Industrial Engineering, University of Iceland, Reykjavik, Iceland
[2]Institute of Physical Geography and Quarternary Geology, Stockholm University, Sweden
[3]Physical Geography, Stockholm University, Stockholm, Sweden
[4]Physical Geography and Ecosystem Analysis, Lund University, Lund, Sweden
[5]Coordination centre for effects (CCE), RIVM, Bilthoven, Netherlands

ABSTRACT

The integrated dynamic ecosystem models SAFE and ForSAFE-VEG were used to estimate critical loads for nitrogen using different chemical and biological criteria, and attempting to account for possible effects of climate change. The study is based on a model application on a set of 640 sites earlier used for critical loads assessments in Sweden, focusing on the ecological effect of nitrogen pollution. To assess critical loads for nitrogen to ecosystems, two criteria were used for the critical loads estimates: 1: N leaching, and 2: changes in plant community composition, and assessed using the models mentioned. We show that the model gives results consistent with earlier results using other methods such as the simple mass balance method. We show that the new model system is operational over the whole of the Swedish territory and may be used for policy development and policy outcome assessments. The critical loads are estimated both under the assumption of no change in climate from today's level and under the assumption that the steady state of climate will be comparable to the predicted levels of 2100 for temperature and precipitation under the IPCC-SRES A2 scenario (Formerly the "worst case" scenario, after 2011, the "medium case" scenario.)

Keywords: ForSAFE-VEG, simple mass balance model, biodiversity, critical loads, exceedence, dynamic modelling

INTRODUCTION

Critical loads of atmospheric deposition were developed to ultimately protect biological endpoints, such as surface waters flora or plant health (Nilsson & Grennfelt 1988, Hettelingh et al. 1992, Warfvinge & Sverdrup 1992). In practice, modelled critical loads and target loads have been estimated based on chemical indicators. Such indicators include, among others, soil solution and surface waters pH and the base cation to aluminium ratio (BC/Al) for acidification, and nitrate concentrations in runoff water for eutrophication. They are meant as proxies to biological impacts of atmospheric deposition on natural ecosystems (Sverdrup et al. 1990, 2007). Already in 1992, critical loads for nitrogen based on ecosystem effects were defined on a conceptual basis (Warfvinge & Sverdrup 1992, Hettelingh et al. 1992) however, integrated models were not really fully up to the task at that time. Chemical proxies were defined for single plant classes and used in a mass balance approach as an approximation until integrated ecosystem effects could be assessed at a later point in time. Empirical critical loads on the other hand have been defined for specific biological endpoints that were assumed to be most relevant for given ecosystems (Achermann & Bobbink 2003, Bobbink & Hettelingh 2011). Both methods (The empirical and the modelled critical loads estimations) have proven valuable within the United Nations' convention on Long Range Transboundary Air Pollution in informing policy about the need to control atmospheric emissions in order to protect ecosystems from adverse effects. As a complement to these methods, recent developments in dynamic modelling of terrestrial ecosystems offer new possibilities for directly relating atmospheric deposition to changes in forest stands and ground vegetation (Sverdrup et al. 2007, 2010, 2011, de Vries et al. 2010, van Dobben et al. 2010, Belyazid et al. 2011). While this new method is still in development by us, it is fully operational and may be able to fill some of the gaps related to the traditional methods.

1. Firstly, dynamic modelling of vegetation changes provides a tool to account for the effects of other drivers, such as climate and land use, simultaneously to the impact from atmospheric deposition, in contrast to empirical critical loads, which consider the effects of atmospheric deposition alone. Thus, not only the final state is assessed, but also the trajectory from the present to the final state. For a system with dynamic changes in the drivers, true steady state may never be fully achieved, and the final state will always be on the way to the next one. Only dynamic models can address such a situation (Sverdrup & Warfvinge 1995).
2. Secondly, it also provides a platform to directly model and quantify the response of specific plants to atmospheric deposition, in contrast to the classical dynamic critical loads' endpoints, which use chemical criteria as proxies for biological impacts. The

chemical proxies often focus on single indicators species, thus eliminating the concept of an ecosystem, as well as ignoring effects of ecological competition. However, early in the critical loads process this was a way to get approximate estimates from a limited information basis (Sverdrup & Warfvinge 1988, Nilsson & Grennfelt 1988, Sverdrup et al. 2007, 2011).

3. Thirdly, whereas the steady state mass balances eliminate time, only looking at the final state at the end of an equilibration time, regardless of far into the future it may be, the dynamic models allow us to assess the whole trajectory through time (Sverdrup & Warfvinge 1995).
4. The new class of models (ForSAFE-VEG, VSD^+-VEG) do for the first time involve ecological competition between species, a prerequisite for being able to say that ecological dynamics is involved. Steady state models and ecological equilibrium models only look at final states and the elements of ecological competition and dynamic change have been removed. These models eliminate the effect of competition and thus the ecological dynamics (Sverdrup et al. 2007, 2011, de Vries et al. 2010, 2013).

Aim and Scope

The aim of this study is to compare critical loads of nitrogen based on the use of chemical indicators (N leaching), and biological indicators (changes in plant community composition) using the integrated terrestrial ecosystem model system called ForSAFE-VEG for 640 forest sites in Sweden (Alveteg et al. 1996, Sverdrup & Warfvinge 1996, Belyazid et al. 2006, Sverdrup et al. 2007. It is outside the scope of this article to describe the databases used, their construction as well as the description of the models and the methodology involved in estimating critical loads for acidity. This study requires as background homework that the reader is familiar with these concepts and models from earlier literature.

METHOD

General Method

The method is to use integrated dynamic ecosystem models to assess critical loads of nitrogen using biodiversity of the ground vegetation as the target function, using regional database of sites used for critical loads for acidity and for forest sustainability assessments and published earlier. The sites are well suited for the purpose as they have complete and connected datasets for these models. For particulars of the datasets used, we recommend to study fundamental concepts described by Hettelingh et al. (1992), Warfvinge et al. (1992), Grennfelt et al. (2001), Alveteg et al. (2004), Belyazid et al. (2004, 2006), Sverdrup et al. (2007, 2008, 2011, 2012a,b). This study is based on dynamic model simulations using the ForSAFE-VEG (Belyaizd 2006) model. ForSAFE-VEG is built around a geochemical core (Belyazid et al. 2004, 2006). ForSAFE-VEG is designed to simulate changes in the composition of the ground vegetation communities (Sverdrup et al. 2007). An extension of the ForSAFE-VEG system to include nutrient dynamics and feedbacks from the ground vegetation layer has been developed (ForSAFE-VEG-GRAFT).

Modelling Vegetation Changes and Estimating Critical Loads at Site Level

A series of dynamic models and methods have been developed over the last years to model the responses of plants to atmospheric deposition and other environmental drivers (de Vries et al. 2010, 2012). The models differ in theory and construction, but are all built with the aim of forecasting the composition of a plant community under given environmental conditions, among which is atmospheric deposition (de Vries et al. 2010). The different models present have been used with different degrees of success at site levels, paving the way for their use in evaluating the effects of different levels of atmospheric deposition on plant communities, and thereby raising the prospect of their use for estimating critical loads. Belyazid and Moldan (2009) used the dynamic model ForSAFE-VEG to simulate ecosystem processes at the well-studied research site Gårdsjön in South Western Sweden, and tested a range of possibilities to obtain critical loads of nitrogen (N) deposition based on changes in the simulated plant community. The study by Belyazid and Moldan (2009) documented three prerequisites to estimate critical loads of N deposition:

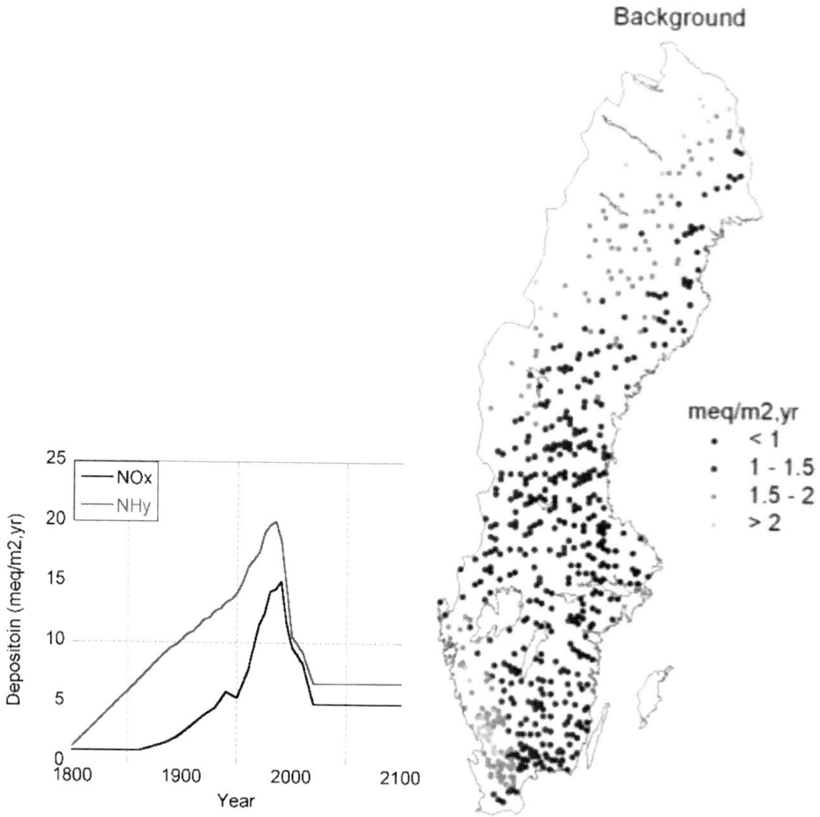

Figure 1. The background deposition used for the assessment, showing shape over time and intensity at different locations.

1. to identify a reference plant community (corresponding to a "clean" reference atmospheric deposition),
2. to select a target population (a subset of the plant community of interest for protection from adverse effects),
3. to define a critical level of unacceptable change due to atmospheric deposition.

Based on a set of clearly defined assumptions regarding the three prerequisites above, Belyazid et al. (2009, 2011a,b) demonstrated the feasibility of this method by estimating preliminary critical loads of N deposition at specific sites in Sweden and Switzerland. Unlike classical chemical indicators, such as base saturation for acidity or nitrogen concentrations for eutrophication, which are one dimensional, using plant communities as indicator requires tracking multiples species simultaneously. The multiple plants occurrences need to be simplified into a single variable that can be tracked over time and on which a limit can be set to identify excessive change. Another difference with the classical critical load method is that instead of defining a fixed critical value (ex. BC/Al = 1), using the plant community as indicator requires a steadily changing reference level, as plant populations are affected by other drivers than nitrogen deposition. To solve this problem, the first step is distilling the relative occurrences of different plants to area cover, and tracking the amount of ground area that is different between different scenarios of N deposition over time. In this way, it is possible to produce a single number over time of the magnitude of the difference in the plant community between two deposition scenarios, expressed in the area that is different between a given scenario and the chosen reference scenario. We decided to refer to the area difference between two plant communities as Mondrians, (M), in reference to the artwork produced by Piet Mondrian (Belyazid 2009, 2010, 2011a,b, 2012). In Mondrian's compositions, each colour in confined to a well-defined area of the canvas, and the entire canvas are occupied by one colour or another. In the same way we model the ground vegetation as species covering a given area. In Mondrian's canvas, if a colour would expand over its given sub-area, it would have to do it at the expanse of another colour on the canvas. In the same way, if a plant expands its cover, it will be at the expanse of other plants at the site which will have to retreat. When a site is simulated under different conditions, the mosaic of plant area cover will look different between different scenarios. The amount of area that is different between two scenarios is referred to as Mondrians, 1M = $0.01m^2$, or 1% of a modelled area. To even out the transient peaks and dips, the cumulative vegetation difference for the period when deposition scenarios differ is divided by the number of years when the deposition scenarios differ to produce a time independent value called the Average Yearly Difference (Belyazid 2010, 2011a,b, 2012).

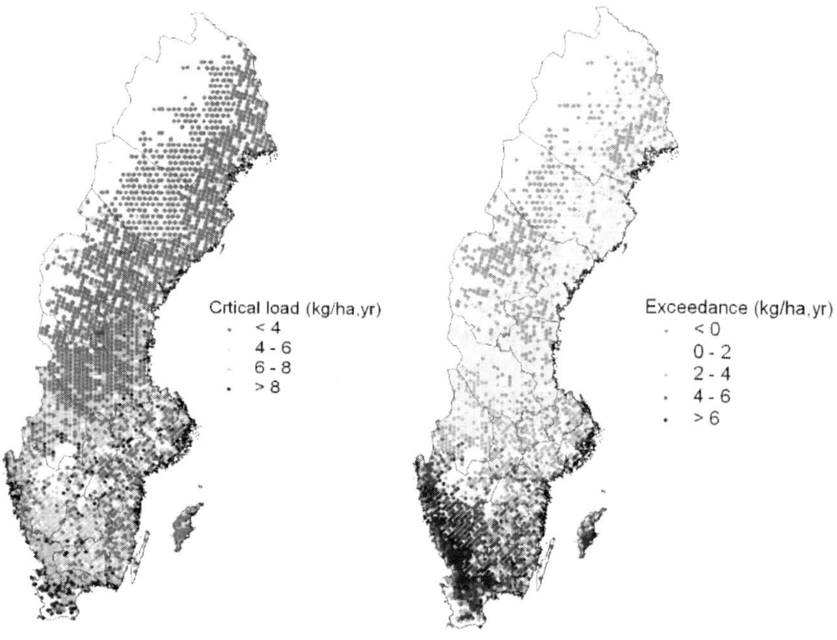

Figure 2. Critical loads (left) and how they are exceeded (exceedance of critical loads) in 2007 (right) as calculated with the simple mass balance model (Akselsson et al. 2007). Simple mass balance model, carried out in 26,763 sites based on the Swedish forest inventory, the forest inventory soil survey, prospecting data of former Terra Mining Company and soil data gained from the Swedish Geological Survey Geochemical mapping Programme.

REGIONAL ESTIMATES OF NITROGEN CRITICAL WITH DIFFERENT CRITERIA

ClnutN Calculated with the SAFE Model System

It has been possible to estimate dynamic critical loads of atmospheric deposition on a regional level, be it national or European. These estimates have formed a cornerstone in informing policy, not only about critical loads but also about target loads. By necessity, the regional dynamic critical loads have so far made use of well-informed generalisations about the inputs they use. For Sweden, Alveteg (1998), Alveteg et al. (2004) and later Belyazid et al. (2004) harmonised inputs from a range of datasets to build an input database for the MAKEDEP and SAFE models, and was able to produce temporally dynamic national estimates of acidification indicators over Sweden. This method was also used in the reporting of dynamic critical loads of acidity for Sweden to CCE as well as forecasts of acidification indicators under different deposition scenarios (Kyrklund et al. 2008). Although this method has proven useable and practical on a regional scale, it is still limited to reproducing site specific dynamic modelling for numerous enough sites to provide a national cover. In essence, the method consists in dynamically modelling specific sites without taking account of possible interactions between the sites. This assumption may be satisfactory when dealing with geochemical indicators, but it may have shortcomings when treating biological indicators as will be discussed below. Based on the database compiled by Alveteg (1998, 2005), critical loads of nutrient N (CLnutN) have been estimated for 640 independent forest sites in Sweden using the SAFE model. The CL calculations were based on the criteria that nitrogen concentrations in runoff water should not exceed 0.3 mg l^{-1} (Figure 1) The map in Figure 2 shows the exceedance of N deposition of the CLnutN level (both in meq $m^{-2}yr^{-1}$) (Adapted after Alveteg 1998, Alveteg et al. 2004, Sverdrup & Warfvinge 1996, Belyazid et al. 2004). The estimated critical loads and their exceedances (Figures 3, 4, 5) are consistent with observations of nitrate leaching from Swedish forest soils (Akselsson et al. 2010). Figure 2 (The Simple mass balance method applied to nirogen critical loads) and Figure 3 (Dynamic assessment using SAFE) show critical loads based on the older methods, applying maximum allowable N concentrations in the percolate, based on empirically set limits (Warfvinge et al. 1992) and Critical loads and exceedance of critical loads as calculated with the simple mass balance model (Akselsson et al. 2007).

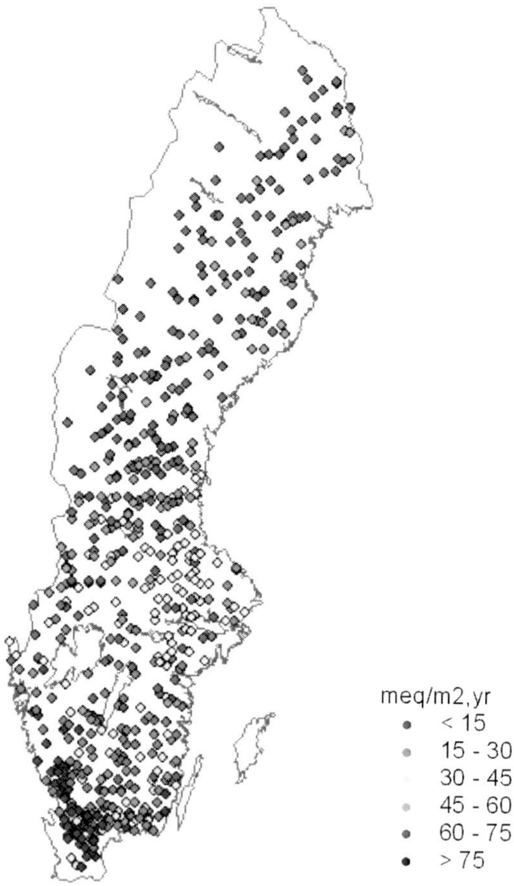

Figure 3. Current legislation deposition exceedance of nutrient N critical loads, calculated with the SAFE model for Sweden (Alveteg 1998, Alveteg et al. 2004, Sverdrup and Warfvinge 1996, Belyazid et al. 2004).

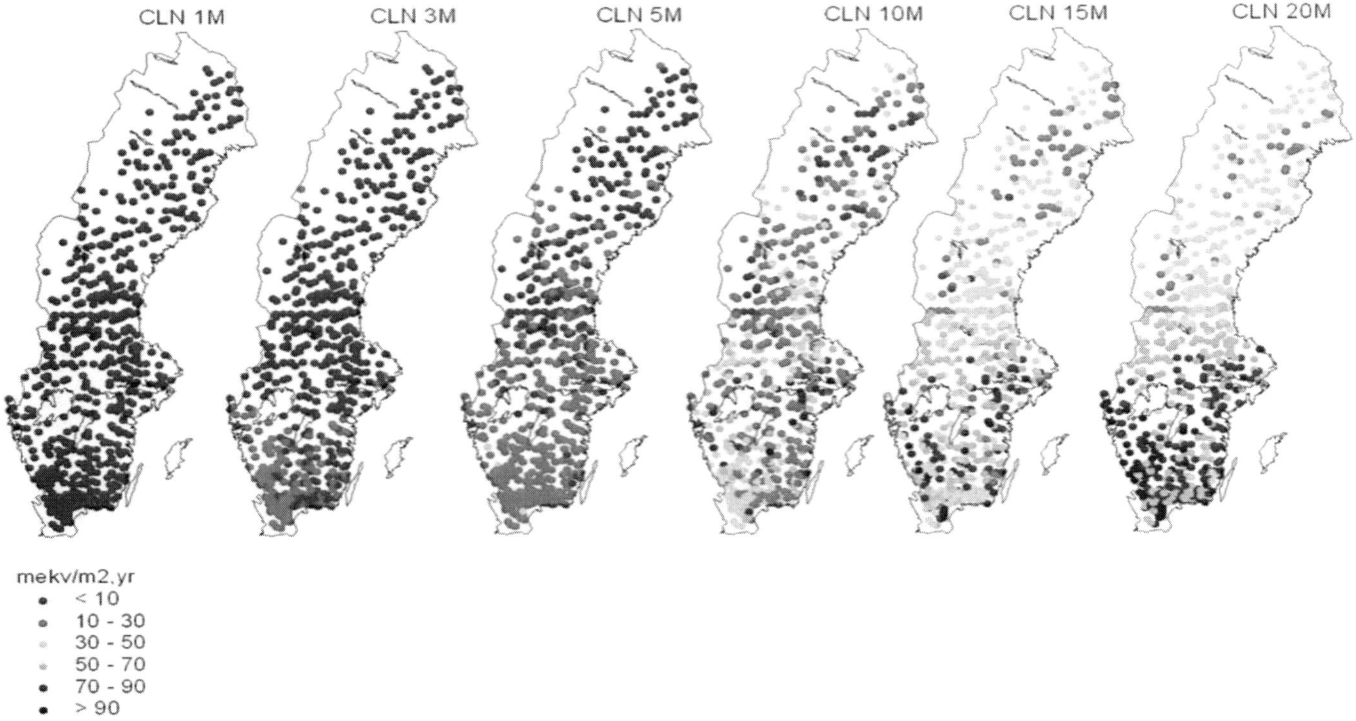

Figure 4. Critical loads for nitrogen, based on a maximum change of 1M (Mondrian) of change, 5M, 10M and 15M based on 42 functional plant classes.

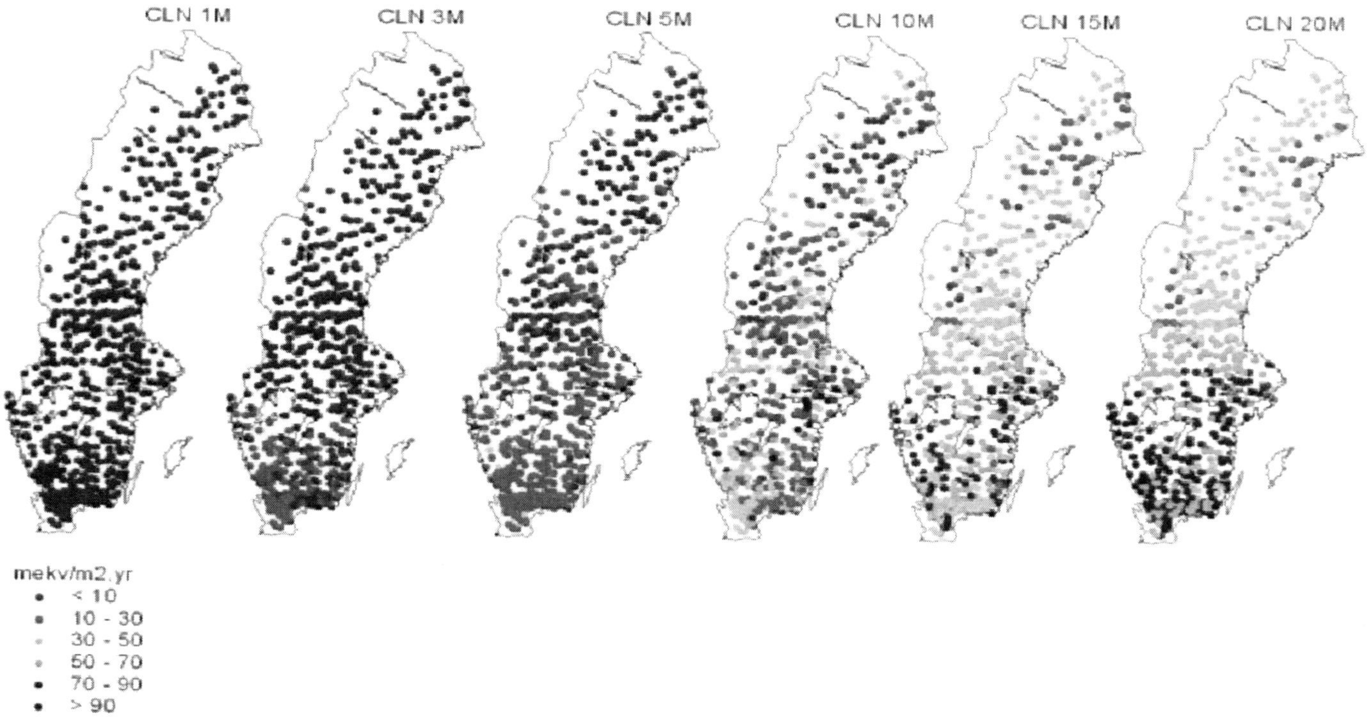

Figure 5. Critical loads for nitrogen, based on a maximum change of 1M (Mondrian) of change, 5M, 10M and 15M based on 362 functional plant classes for Europa.

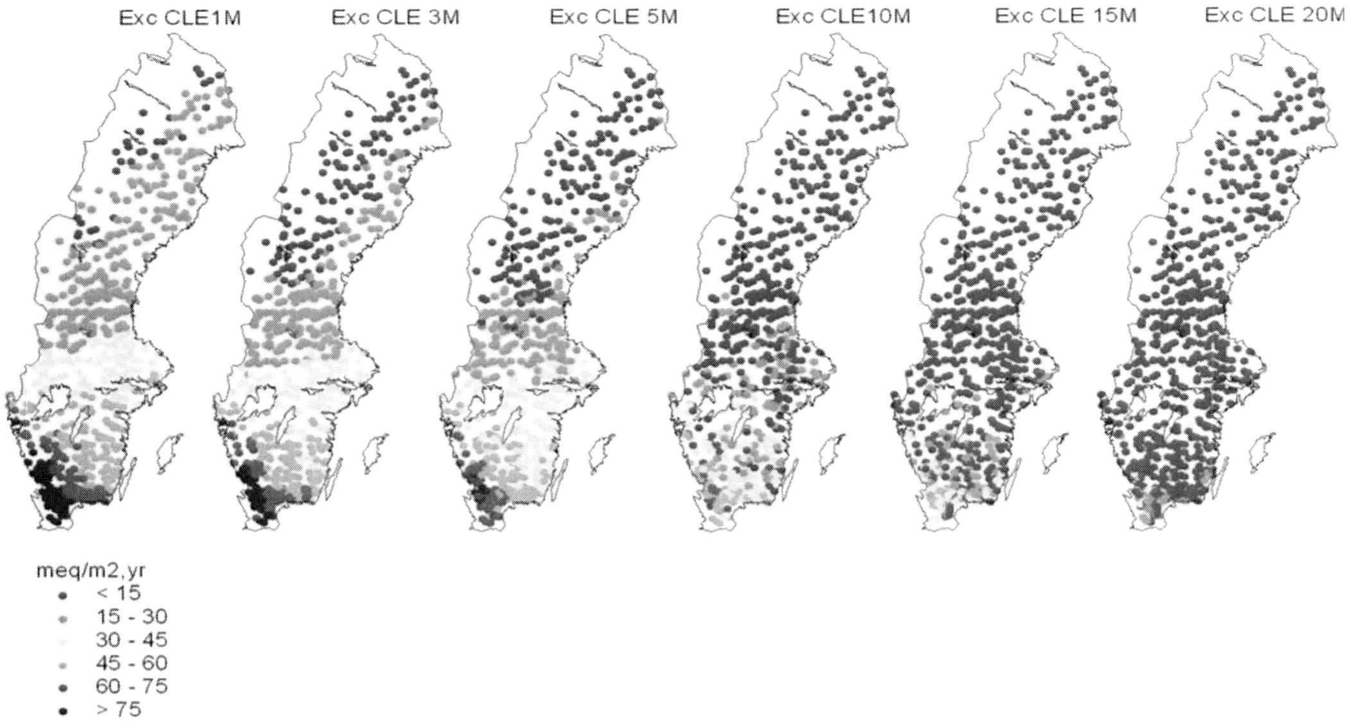

Figure 6. N exceedances of CLE N deposition over the CLvegN causing at most 1M (Mondrian) of change, 5M, 10M and 15M. One Mondrian unit in this case is equivalent to 1% of area cover by plants. The ForSAFE-VEG model was used 42 plant groups.

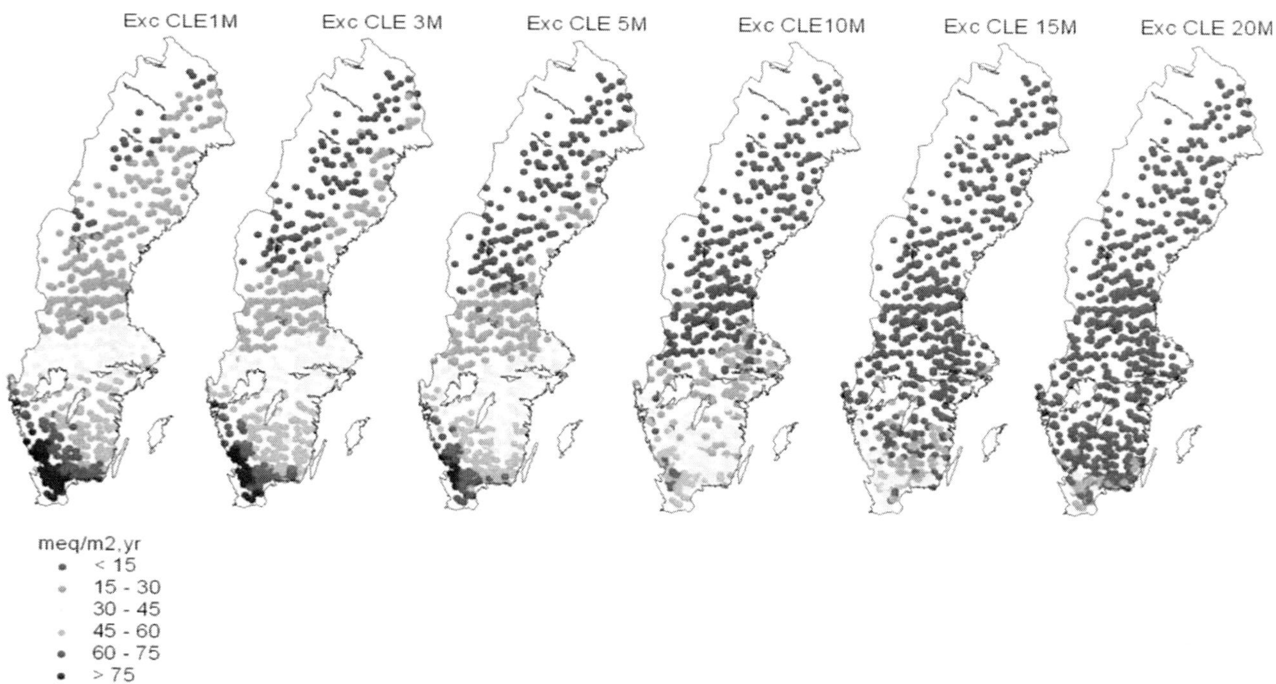

Figure 7. N exceedances of CLE N deposition over the CLvegN causing at most 1M (Mondrian) of change, 5M, 10M and 15M. One Mondrian unit in this case is equivalent to 1% of area cover by plants. The ForSAFE-VEG model was used 362 plant groups.

CLvegN with ForSAFE-VEG

The same sites used in Figures 2 and 3 have been simulated with the ForSAFE-VEG model (Belyazid 2006, Belyazid et al. 2010, 2011a,b, 2012, Sverdrup et al. 2007, 2008, 2010, 2012), this time with focus on changes in the composition of the ground vegetation communities rather than N leaching (Figure 4). The model estimated the critical loads of N deposition with the assumption that any change in any constituent of the plant community is proportional in importance to the reference occupancy of that plant, meaning that a change in a marginal plant is reflected as marginal, while a change in a dominant plant can be more strongly reflected in the CL estimation.

With the acknowledge limitations of this method of weighing the plants in a community (van Dobben et al. 2010), the model can produce sensible results (Figure 4). Depending on the critical limit of change which is tolerated for the critical load (i.e., the level of change which is still below a significant harmful effect), different critical loads are produced, and the lower the critical limit (the more stringent the definition of a harmful effect is), the lower the critical load (the lower deposition level is needed to cause the harmful effect). Similarly to the exceedances of critical loads presented for nutrient N, Figure 7 shows the exceedances of CLE deposition to different critical loads defined for gradually higher critical limits of unacceptable change (i.e., increasingly more tolerant definitions of the level of harmful effect). With a critical limit of 1M, CLE deposition exceeds the critical load in nearly all of Sweden, with the exception of some of the mountainous ecosystems in the northeast of the country. A critical limit of 1M in this case corresponds to a change of 1% of the total area cover of a plant community (for a more thorough discussion of the Mondrian (M) unit, see Sverdrup et al. (2008) and Belyazid et al. (2010, 2011a,b). On the other side of the tested scale, CLE deposition exceeds the critical load of N only in parts of southern and south-western Sweden when the critical limit is set to 15M.

Comparing Estimates of CLnutN and CLvegN

Figure 5 shows the critical loads and exceedance of critical loads as calculated with the simple mass balance model (Akselsson et al. 2007), the model used before biodiversity-based methods were available. The Simple mass balance model as it is called was carried out in 26,763 sites across Sweden. The critical load of N estimated based on nutrient N in runoff water (CLnutN) is comparable to that estimated at 10M (CLvegN) (i.e., equivalent to 10% loss of plant community similarity) based on changes in plant communities. Roughly, this comparison means that if atmospheric deposition of N is not allowed to exceed the level causing runoff N concentrations to reach or exceed 0.3 mg N l^{-1}, it will not cause more than 10M changes in forest floor plant communities. The opposite is also true. The main difference between the two methods remains the fact that CLnutN relates to a chemical indicator, namely N concentrations in runoff, which can have different effects on biological elements of the environment depending on the state of other abiotic conditions. Nutrient N is a strong driver in Swedish ecosystem due to the fact that it has been one of the most prominent limiting nutrients in these ecosystems. Yet, due to the on-going climatic changes and changes in land use, other environmental parameters may take precedence over N in steering biological developments in boreal ecosystems (Sala et al. 2000). In contrast, the method estimating CLvegN takes into account as many environmental drivers as understood, including land uses and climatic trends. The dynamic models used to simulate vegetation changes as part of the wider ecosystem (in the case of this study it is the ForSAFE-Veg model), are able to integrate multiple dynamic drivers simultaneously over time.

In this manner, the CLvegN method is able to distil the exclusive effect of atmospheric deposition. The key to doing this is that the dynamic models allow for a revision of the steady state assumption inherent in the critical loads concept. A central postulation that makes the use of chemical proxies for biological impacts valid is the assumption of steady state. This assumption states that ecosystems are subject to unchanging environmental conditions, and under these conditions it is possible to find a critical level of atmospheric deposition that causes unwanted change in specific ecosystem endpoints. What this assumption implies is that if all drivers are "constant" over time with the exception of deposition, it is possible to simulate a number of scenarios of deposition, allowing us to find the specific level of deposition that causes the harmful effect of interest (since deposition is the only changing driver). In this reasoning, if the term "constant" is replaced with "similar", it still remains theoretically possible to extract the exclusive effect of deposition (since all other drivers are similar among the scenarios evaluated). The implication of such a modification is that instead of finding a steady state critical load, the method now produces a critical load within a fixed time frame, i.e., instead of CLnutN never causing N leaching to exceed 0.3 mg N l^{-1} (compare those with Warfvinge & Sverdrup 1992 and Warfvinge et al. 1992), CLvegN will not cause plant communities to change by more than 10M by the year 2100. At the same time, any change brought about by any driver other than deposition will still be allowed for (as for changes in plant communities due to temperature increases associated with the on-going climate change).

DISCUSSION

Choosing a target plant community, or a subset of a community, is central to the outcome of the CLvegN estimates (Belyazid & Moldan 2009). Yet, there is no consensus among ecosystem specialist as to which segment of a plant community is to be protected from adverse effects due to N deposition. This is logical given that biomes are widely different, but also because the purpose of conservation or protections can vary vastly. Van Dobben et al. (2010) provide an inclusive summary of the possible criteria to be used for weighting specific plants in a community to provide an index of biodiversity status. The study of van Dobben et al. (2010) lists five possible criteria for weighting which we have elaborated:

1. no weighting (i.e., all plants are equally important),
2. weighting according to plant abundance (such as the Simpson or the Shannon-Weaver indices, this is also the weighting method used in this study),
3. Involving value statements and policy considerations;
4. Weighing to the intrinsic value of a species (such as endemic or red listed species), an environmental or natural conservation policy value judgement must be made.
5. weighting to the desirability of a species, an environmental or natural conservation policy judgement must be made.
6. weighting to functional groups. A definition of the functionality must be made, which may potentially involve value statements based in environmental or natural conservation considerations

Choosing between these options goes back to setting a value on the aspect of plant communities to be protected within a given ecosystem. While the answer to the question: "what should be protected?" remains beyond the scope of this paper, such a methodical approach for weighting plants is needed when regionalising the CLvegN estimations. The CLvegN estimations are based on the impact of N deposition on plant communities at specific sites, without consideration of the surrounding biotas. This is probably the biggest shortcoming of the method presented above using biological endpoints to estimate the critical loads of deposition. Unlike with chemical endpoints (as in the case of CLnutN), it is imperative to have a clear picture of the ecology surrounding each modelled site in order to appropriate represent seed dispersion and plant colonization. This is not in place I any of the models used at present. It can be argued that in the context of critical loads (i.e., in the very far future) plants will have enough time to colonize a given site even if today it lies far from their geographical reach.

This study illustrates that the methods for estimating critical loads for nitrogen with respect to ecological change has improved a lot over the last 20 years, to a point where we may actually validate the results on field observations. When playing with the model, and the basic way to set it up for assessment, it becomes apparent that the number of plant groups used, is not of great importance for the outcome. Using 43 plant groups or the whole European plant group list of 363 groups yielded for practical policy purposes the same result. This shows that there is a large degree of ecological redundancy among the plant groups, and 43 groups seem like sufficient to cover the whole range of ecological niches. Thus, not all the plant species need to be part of the model, but we may relate the assessments to the full list as long as we are able to address each of the 363 plant groups to one of the 43 aggregated plant groups used in the assessment. In the same way, the accuracy of the predicted changes increase as we add plant groups, but after including about 30 groups, the performance starts to decline (Sverdrup et al. 2010). The latest versions of the FarSAFE-VEG model are operationally ready for regional assessments of critical loads for acidification and nitrogen with respect to ecological effects such as ground vegetation biodiversity being used as the objective function. When used properly, the results are verifiable on field observations and yield robust policy support assessments. We have shown that this method is backwards compatible and well as it expresses different degrees of ecological impact through the new Mondrian parameter unit.

CONCLUSION

We may make some conclusions:

1. The ForSAFE-VEG model system is adequate and operational for assessing critical loads for nitrogen based on biodiversity criteria
2. A sufficient parameterization exists for ground vegetation for making an assessment. We do not need to include all plant species, 30-50 major plant groups will suffice for the assessment. All other species can be linked to those functional groups
3. The outcomes are consistent with earlier assessments

Acknowledgments

The work in this study was made possible with grants from MISTRA (ASTA I+II and SUFOR I+II research programmes), from the Swedish Environmental Protection Board in Stockholm (The Swedish Critical Loads Programme), the German Ministry of Environment (UBA) in Berlin, the Swiss Federal Ministry of Environment and landscapes (FOEN) in Bern and K. A. Rasmussen Precious Metals Industries a/s, Hamar, Norway.

Conflict of Interest

Author confirms that this article content has no conflict of interest.

References

Achermann, B & Bobbink, R (Eds) (2003) *Empirical Critical Loads for Nitrogen*. Environmental Documentation No. 164. Air. Swiss Agency for Environment, Forest and Landscape SAEFL, Berne.

Akselsson, C, Westling, O, Sverdrup & H, Gundersen, P (2007) Nutrient and carbon budgets in forest soils as decision support in sustainable forest management. *Forest Ecology and Management,* 238, 167-174.

Akselsson, C., Sverdrup, H. & Holmqvist, J (2005) Estimating weathering rates of Swedish forest soils in different scales, using the PROFILE model. *Journal of Sustainable Forestry*, 21, 119-131.

Akselsson, C, Westling, O, Sverdrup, H, Holmqvist, J, Thelin, G, Uggla, E & Malm, G (2007) Impact of harvest intensity on long-term base cation budgets in Swedish forest soils. *Water, Air and Soil Pollution: Focus,* 7, 201-210.

Akselsson, C, Belyazid, S, Hellsten, S, Klarqvist, M, Pihl-Karlsson, G, Karlsson, PE & Lundin, L (2010) Assessing the risk of N leaching from Swedish forest soils across a steep N deposition gradient in Sweden. *Environmental Pollution,* 158, 3588-3595.

Alveteg, M (1998) Dynamics of forest soil chemistry. PhD dissertation. Lund University, Lund.

Alveteg, M (2004) Projecting regional patterns of future soil chemistry status in swedish forests using SAFE. *Water, Air and Soil Pollution: Focus,* 4, 49-59.

Alveteg, M, Sverdrup, H & Warfvinge, P (1996) Regional assessment of dynamic aspects of soil acidification in southern Sweden. *Water, Air and Soil Pollution,* 85, 2509-2514.

Belyazid, S (2006) Dynamic modelling of biogeochemical processes in forest ecosystems, Ph.D. thesis, Lund University.

Belyazid, S & Moldan, F (2009) Using ForSAFE-Veg to investigate the feasibility and requirements of setting critical loads for N based on vegetation change — pilot study at Gårdsjön. Report to Naturvårdsverket, 28p.

Belyazid, S, Sverdrup, H & Alveteg, M (2004) Biogeochemical models family of the biogeochemistry group at Lund University. Bashkin, V (Ed). Proceedings of the 6th Subregional Meeting on Effect-Oriented Activities in the Countries of Eastern and South-Eastern Europe Relevant for the UN ECE Convention of LRTAP. Moscow-Puschino, Russia, 12-28.

Belyazid, S, Sverdrup, H, Kurz, D, Braun, S & Rhim, B (2009) Estimating critical loads of nitrogen deposition using the composition of the ground vegetation as a biological indicator, report to the CCE. Swedish Environmental Protection Agency and the Swiss Federal Office for the Environment.

Belyazid, S, Bailey S & Sverdrup, H (2010) Past and future effects of atmospheric deposition on the forest ecosystem at the Hubbard Brook Experimental Forest: simulations with the Dynamic Model ForSAFE In: Hanrahan. G (Ed). Modelling of pollutants in complex environmental systems, Volume II: 357-377, ILM Publications, International Labmate Limited.

Belyazid, S, Sverdrup, H, Kurz, D & Braun, S (2011a) Exploring ground vegetation change for different scenarios and methods for estimating critical loads for biodiversity using the ForSAFE-VEG model in Switzerland and Sweden. *Water, Air and Soil Pollution*, 216, 289-317.

Belyazid, S, Kurz, D, Braun, S, Sverdrup, H, Rihm, B & Hettelingh, JP (2011b) A dynamic modelling approach for estimating critical loads of nitrogen based on plant community changes under a changing climate. *Environmental pollution*, 159, 789-801.

Belyazid, S, Kurz, D, Sverdrup, H, Braun, S & Rihm, B (2012) Developing a method for estimating Critical Loads of Nitrogen deposition based on biological indicators under a changing climate. In: de Vries, W (Ed) Critical loads in Europe. Springer Verlag.

Bobbink, R, Hettelingh, JP (Eds) (2010) Review and revision of empirical critical loads. National Institute for Public Health and the Environment, the Netherlands.

Dobben van, H, Hettelingh, JP, de Vries, W, Slootweg, J & Reinds, GJ (2010) Plant species diversity indicators for impacts of nitrogen and acidity and methods for their simulation: an overview. In: Slootweg, J, Posch, M and Hettelingh, JP (Eds). CCE status report 2010.

Grennfelt, P, Moldan, F, Alveteg, M, Warfvinge, P & Sverdrup, H (2001) Critical loads - Is there a need for a new concept. *Water, Air and Soil Pollution Focus,* 1, 21-27.

Hettelingh, JP, Posch, M, de Vries, W, Bull, K & Sverdrup, H (1992) Guidelines for the computation and mapping of nitrogen critical loads and exceedance in Europe. In: Grennfelt, P-I & Lövblad, G. (Eds) Critical loads for nitrogen — A workshop report, 287–303. Nordic Council of Ministers.

Jönsson, C, Warfvinge, P & Sverdrup, H (1993) The SAFE model applied to the Solling dataset. *Ecological Modelling*, 83, 85–96.

Martinson, L, Alveteg, M, Kronnäs, V, Sverdrup, H, Westling, O & Warfvinge, P (2005) A regional perspective on present and future soil chemistry at 16 Swedish forest sites, *Water, Air, and Soil Pollution,* 162, 89-105.

Nilsson, J & Grennfelt, P (Eds) (1988) Critical loads for sulphur and nitrogen. Report 1988:15. Nordic Council of Ministers, Copenhagen, Denmark.

Sala, OE, Chapin FS, Armesto, JJ, Berlow, R, Bloomfield, J, Dirzo, R, Huber-Sanwald, E, Huenneke, LF, Jackson, RB, Kinzig, A, Leemans, R, Lodge, D, Mooney, HA, Oesterheld, M, Poff, NL, Sykes, MT, Walker, M, Walker, BH & Wall, DH (2000) Global biodiversity scenarios for the year 2100. *Science,* 287, 1770-1774, dx.doi.org/10.1126/science.287.5459.1770.

Sala, A, Piper, F & Hoch, G (2010) Physiological mechanisms of drought-induced tree mortality are far from being resolved. *New Phytol*, 186, 274–281.

Sverdrup, H & Warfvinge, P (1995) Past and future changes in soil acidity and implications for forest growth under different deposition scenarios. *Ecological Bulletins*, 44, 335–351.

Sverdrup, H, Warfvinge, P, Blake, L & Goulding, K (1995) Modeling recent and historic soil data from the Rothamsted Experimental Station UK, using SAFE. *Agriculture, Ecosystems and Environment*, 53, 161–177.

Sverdrup, H, Warfvinge, P & Jönsson, C (1993) Critical loads of acidity and nitrogen for forest soils, groundwater and first order streams in Sweden. In: Hornung, M and Skeffington, R (Eds) Critical loads, Concepts and Applications, 54–67. UN/ECE workshop at Grange-over-Sands, Cumbria. HMSO.

Sverdrup, H & Rosen K (1998) Long-term base cation mass balances for Swedish forests and the concept of sustainability. *Forest Ecology and Management,* 110, 221-236.

Sverdrup, H, Warfvinge, P & von Brömssen, U (1987) A mathematical model for podzolic soil profiles exposed to acidic deposition. In: Christophersen, N (Ed). *Acidification and Water Pathways*, 435-444, ISBN 82-554-0486-4. Norwegian National Committee for Hydrology

Sverdrup, H, Belyazid, S, Nihlgård, B & Ericson, L (2007) Modelling change in ground vegetation response to acid and nitrogen pollution, climate change and forest management at in Sweden 1500-2100 AD. *Water, Air and Soil Pollution Focus,* 7, 163-179.

Sverdrup, H, Belyazid, S, Kurz, D & Braun, S (2008) *A proposed method for estimating critical loads for nitrogen based on biodiversity using a fully integrated dynamic model, with testing in Switzerland and Sweden.* In: Sverdrup, H (Ed.). Towards critical loads for nitrogen based on biodiversity. Background document for the 18th CCE workshop on the assessment of nitrogen effects under the ICP for Modelling and Mapping, LRTAP Convention (UNECE), Bern, Switzerland, 21-25 April 2008, 3-37. Swiss Federal Office of the Environment (FOEN), Bern, Switzerland.

Sverdrup, H, Nihlgård, B & Belyazid, S (2011) Mapping critical loads for nitrogen based on biodiversity using ForSAFE-VEG: Introducing the basic principles In: Sutton, M, Mason, K, Sverdrup, H and Sheppherd, L (Eds,) Proceedings of the Conference on Reactive Nitrogen at Edinburgh, September 2009, 1-6.

Sverdrup, H, Belyazid, S, Kurz, D & Braun, S (2012a) *Use of an integrated deterministic soil-vegetation model to assess impacts of atmospheric deposition and climate change on plant species diversity.* In: de Vries, W and Hettelingh, JP (Eds). Critical Loads for Nitrogen, Acidity and Metals for Terrestrial and Aquatic Ecosystems. Springer Verlag.

Sverdrup, H, McDonnell, TC, Sullivan, TJ, Nihlgård, B, Belyazid, S, Rihm, B, Porter, E, Bowman, WD & Geiser, L (2012b) Testing the feasibility of using the ForSAFE-VEG model to map the critical load of nitrogen to protect plant biodiversity in the Rocky Mountains region, USA. *Water, Air and Soil Pollution*, 223, 371-387.

Vries, de W, Posch, M, Sverdrup, H and de Wit, H (2013). Chemical indicators for use in risk assessments of inputs of nitrogen, acidity and heavy metals. In: de Vries, W and Hettelingh, JP (Eds). *Critical Loads for Nitrogen, Acidity and Metals for Terrestrial and Aquatic Ecosystems*. Springer Verlag.

De Vries, W, Wamelink, GWW, van Dobben, HF, Kros, J, Reinds, GJ, Mol-Dijkstra, JP, Smart, SM, Evans, DR, Rowe, EC, Belyazid, S, Sverdrup, H, Hinsberg, A van, Posch, M, Hettelingh, JP, Spranger, T & Bobbink, R (2010) Use of dynamic soil–vegetation models to assess impacts of nitrogen deposition on plant species composition: an overview. *Ecological Applications,* 20, 60-79.

Warfvinge, P, Sverdrup, H & Rosen, K (1992) Calculating critical loads for N to forest soils. In: Grennfelt PI and Lövblad, G (Eds) Critical loads and levels for nitrogen. *Nordic Council of Ministers. Nord* 1992, 41, 403–417.

Warfvinge, P & Sverdrup, H (1992) Calculating critical loads of acid deposition with PROFILE – a steady–state soil chemistry model. *Water, Air and Soil Pollution*, 63,119–143.

Warfvinge, P, Falkengren-Grerup, U, Sverdrup, H & Andersen, B (1993) Modeling long–term cation supply in acidified forest stands. *Environmental Pollution*, 80, 209–221.

Warfvinge, P, Sverdrup, H, Alveteg, M & Rietz, F (1996) Modelling geochemistry and lake pH since glaciation at lake Gårdsjön. *Water, Air and Soil Pollution,* 85, 713-718.

In: Ecological and Biogeochemical Cycling … ISBN: 978-1-53612-081-3
Editor: Vladimir N. Bashkin

Chapter 17

ENVIRONMENTAL RATINGS AS A FACTOR OF IMPROVING INVESTMENT ATTRACTIVENESS OF RUSSIAN OIL AND GAS COMPANIES OPERATING IN THE ARCTIC

***Olga P. Trubitsina*[*] *and Vladimir N. Bashkin*[2,3]**

[1]Northern (Arctic) Federal University,
North Dvina Bank, Russian Federation
[2]Institute of Physicochemical and Biological Problems of Soil Sciences of Russian Academy of Sciences, Pushchino, Russian Federation
[3]Institute of Natural Gases and Gas Technologies – Gasprom VNIIGAZ LLC, Razvilka, Russian Federation

[*] Corresponding author: Northern (Arctic) Federal University, North Dvina Bank, 17, Arkhangelsk, 163002, Russian Federation, tel +79116709225, email: test79@yandex.ru.

ABSTRACT

The article is devoted to the issues of environmental ratings as a factor of improving investment attractiveness of Russian oil and gas companies, operating in the Arctic. The authors disclose the need of using environmental rating for oil and gas industry in Russia. Particular attention is given to the issues of rating results of Environmental Responsibility of Oil and Gas companies in Russia that was first held in 2014 and then in 2015. It was conducted by the cooperative initiative by CREON Group and WWF Russia with participation of National Rating Agency. The authors have selected from all Russian oil and gas companies only those who operating in the Arctic region and these have been analyzed.The rating's results show that there are "top three" among russian oil and gas companies, operating in the Arctic. They are Gazprom, Sakhalin Energy (Sakhalin-2), Zarubezhneft. There are also different companies in rating's sectors. For instance, the best position of environmental management belong to Sakhalin Energy (Sakhalin-2), Gazprom, Rosneft. In whole, the environmental rating of the russian oil and gas companies, operating in the Arctic, may create a new mechanism for informing foreign investors, which may leads to the rise of FDI in Russia and lowering of the GER (geoenvironmental risk) in conditions of Arctic's exploitation accordingly.

Keywords: environmental rating, oil and gas industry, geoenvironmental risk, investment attractiveness, the Arctic

1. INTRODUCTION

It's believed with today's energy-hungry way of life, the need for oil is at an all-time high (Improving safety 2014). According to the IEA's World Energy Outlook, global energy demand will grow by more than a third over the period to 2035 (Bigliani 2013). With this surge in energy demand, the oil and gas industry is being challenged to improve production efficiencies and to maintain a growth trajectory in its crude oil and natural gas resources. The ability of the industry to address a number of operational challenges will be critical in ensuring that future demand is met with adequate supply (Improving safety 2014). Human and environment safety and health protection remains the number 1 priority for the oil and gas industry (Bigliani 2013).

The Russia possesses one-third of the world's known natural gas reserves and is one of the major oil-producing countries in the world acording to the International Energy Agency. (Wernstedt 2002). The Arctic constitutes an area of new opportunities along with the geoecological and geopolitical challenges for oil and gas companies. The region is expected to play a vital role in meeting Russia's energy supply over the next several decades that is why it is important to outline the main problems and find ways for improvement in the oil and gas sector. The companies that operate in the Arctic should be analysed in the sphere of ecological awareness of the consequences that might be faced when oil and gas are exploited there.

The activity in the oil and gas industry in the Arctic Zone of the Russian Federation (AZRF) has been on the increase for the past several years and the geoenvironmental risks (GER) have been growing accordingly. In comparison to many other parts of Russia, the Russian Arctic is considered to be a relatively pristine area. However, "contamination and other negative effects in the AZRF result in the formation of "hot spots" and "impact areas" characterized by a high level of chemical contamination of the environment and transformation of the natural geochemical background (Diagnostic analysis 2011, Trubitsina 2016). There is always a variety of potential GER and negative impact on the environment because of oil and gas exploration. Consequences depend upon the stage of the process, the size and complexity of the project, the nature and sensitivity of the surrounding environment and the effectiveness of planning, pollution prevention, mitigation and control techniques (Environmental Management 1997). From all variety of effects, for example, influencing the biota at placing marine pipelines in the Arctic, we can distinguish two groups of factors which differ by the nature of their effects on natural systems: operation of mechanisms and processes connected thereof (increased water turbidity due to high concentrations of a suspended bottom sludge at excavation of trenches in shallow areas), and hydrocarbon effects (Trubitsina 2016). Thus the GER indicators are defined as risks which occur in the environmental and climatic conditions of the Arctic in the "industry - environment" system, as related to the mutual impacts of industrial facilities on the environment and of the environment on industrial facilities (Bashkin 2014, Trubitsina & Bashkin 2016).

At the same time russian oil and gas industry also is faced with barriers to foreign direct investment (FDI) arising from the uncertainty of the existing legal regime of Arctic (geopolitic issues) and the readiness of technologies to cope with geoecological challenges. This underinvestment may lead to oil spills and pipeline leaks that have influence on increasing GER. This article focuses on certain points of methods, used to assess the outside of investment appeal of a russian oil and gas companies, operating in the Arctic, in particular the use of existing assessment results, provided by credit rating agencies, leading journals, etc. The article is devoted to the issues of environmental rating as a factor of improving outside investment attractiveness of companies through raising the environmental awareness of business about activity connected with striving towards the reduction of GER and sustainable development of the Arctic region.

2. Why Environmental Rating is Needed for Russian Oil and Gas Companies?

The United Nations Conference on Environment and Development (UNCED) held in Rio de Janeiro in June 1992 - 'The Earth Summit'- focused world attention on the close links that exist between the environment and socioeconomic development (Environmental Management 1997). According to the UNCTAD in order to achieve the intended goal of the sustainable development the "international system of investment management needs to be reformed" (Global Investment 2015, Trubitsina & Bashkin 2016). Global flows of FDI rose in 2015 (+36%) to reach an estimated US$1.7 trillion, their highest level since 2007. However, this figure changes non-uniformly in different countries. Thus, while FDI has increased in countries with developed economies including the EU and the United States by 90% compared to the last year, it also showed a 92% drop in Russia, which is linked to regional tensions, elevated GERs and reduced market confidence (Global Investment 2015).

Stimulation of investment activities in Russia under current conditions is instrumental for increasing the national economic activity, integration into global economy and financial market (Strizhova 2015). Sectors of economy related to exploitation of natural resources and having a good export potential (including oil and gas industry) (Official site 2016) are seen by foreign partners as the most attractive sectors of economy for investing. Therefore, voluntary submission of reports as part of raising the environmental awareness of business are paramount in boosting the investment appeal of enterprises in this industry. Quantitative assessment of investment appeal of an enterprise presumes that the environmental performance stands out among other supposed parameters and that it is assessed mainly in terms of fines for environmental law violations. The generalizing factor is calculated as a ratio of fines for environmental law violations to net profit. Zero shall be preferred value for this figure. However, it is no less important to take into account conservation and environmental protection activities of an enterprise as part of its short-term and long-term planning. The interaction of economics and ecology brings up new vision of cost estimate and internalization, i.e., the need to consider external impact on the environment in enterprises' financial reports (Trubitsina & Bashkin 2016).

Contemporary environmental regulatory tools draw on a combination of 'hard' mandatory regulatory measures imposed by the government and 'soft' voluntary commitments to environmental responsibility standards adopted by companies. Mandatory regulatory requirements obstruct attempts by companies to minimize costs associated with environmental responsibility standards, in order to force them to meet at least the market average standard. At the same time, some industries choose to adopt voluntary environmental standards, which they believe will increase their competitive advantage in environmentally sensitive markets and afford access to longer-term and affordable sources of finance (Beurden &

Gossling 2008, Stuebs & Sun 2010, Prakash & Potoski 2012, Shvarts et al. 2016, Sharkey & Bromley 2015).

Environmental ratings affect the market, concerned in financial outcome, in such a way, that they provide unbiased evaluation of technologic aspect of production. The new service market is currently emerging in Russia - providing of information on environmental performance of companies and enterprises. As the lack of unbiased data in this field is made up for, it will allow for more effective application of environmental criteria to investment, equipment, goods, and service markets (Avramenko et al. 2006).

Late 1990s marked the development of environmental ratings as part of raising the environmental awareness of business strategy in order to find the best in terms of their investment appeal. So in 1997 the Global Reporting Initiative (GRI) was established with the trinity of economic, social, and environmental performance of a company as its manifest. This is the voluntary international standard for companies, which report on sustainable development. However, unlike ISO 14000 compliance certificates, which are all but mandatory for companies with big claims for good market positions, the GRI is aimed more at increasing the investors' confidence in reporting companies, improving their goodwill, and cost reduction through new business opportunities. GRI is not yet common in Russia; however, some of the largest Russian corporations perceive participation in World Business Council For Sustainable Development as a key instrument in promoting their influence on the global market (Trubitsina 2014).

The introduction of environmental ratings make it possible to prioritize attracting of investments to those russian oil and gas companies, which the most meet the sustainable development criteria. The growing interest of investors (both institutional and private) towards ratings provokes companies to view them as a drive to continuously improve their internal environment (Trubitsina & Bashkin 2016).

3. Environmental Ratings of Russian Oil and Gas Companies, Operating in Arctic

Ratings and ranking by third-party evaluators have been becoming increasingly important as tools for improving corporate environmental performance. An increase in the amount of peer companies being rated is often associated with emissions reductions (Shvarts et al. 2016, Sharkey & Bromley 2015).

3.1. Rating Objectives

A good tool for improving outside investment attractiveness of russian oil and gas companies, operating in the Arctic is a environmental rating. Authors describe this issue according to rating of Environmental Responsibility of Oil and Gas companies in Russia that was first held in 2014 and then in 2015. It was conducted by the cooperative initiative by CREON Group and WWF Russia with participation of National Rating Agency (Sharkey & Bromley 2015, Knizhnikov et al. 2015).

The authors aimed at provide an unbiased and comparable data on environmental responsibility of participants and the impact of russian oil and gas companies on the Arctic's environment. It is believed that investors should have more excess to such information that will influence the reputation of russian oil and gas industry players, operating in the Arctic and, ultimately, promote the removal of barriers for FDI and decrease GER there.

3.2. The Rating Method

The authors have selected from all russian oil and gas companies only those who operating in the Arctic region and they have analyzed of companies' activities in three sections in accordance with (Shvarts et al.

2016). The first one is environmental management. In most cases criteria in this section are more rigid compared to Russian legislation on environmental protection. These criteria correspond to the best global standards and practices in the oil and gas business. The second one is environmental impact. Section evaluates the scale of oil and gas companies' impact on the environment. In particular, the level of damage is assessed for air, water, and land resources during project implementation. In most cases the criteria are based on components of official statistics on environmental protection. The third one is disclosure/transparency. Section evaluates the extent of companies' readiness to disclose information with respect to the environmental impact of their industrial activities (Knizhnikov et al. 2015). The rating's ecological criteria are based on (Environmental Standards for Operations 2005), a document developed by a coalition of non-governmental environmental organizations as a common environmental public standard for improvement of oil and gas companies' environmental policies, procedures and practices. Criteria of each sections for rated companies are showed in Table 1.

Table 1. Rating's criteria for oil and gas companies, operating in the Arctic in accordance to data material (Shvarts et al. 2016, Knizhnikov et al. 2015)

Environmental rating includes criteria, grouped into three sections
Section 1. Environmental management
1. Presence of an Environmental Management System certified in accordance with the ISO 14001 standard; 2. Inclusion of certain requirements, bans, and commitments in the company's Environmental Policy and publicly available company documents; 3. Availability of a company policy or any other official company document on engagement with indigenous peoples of the Russian North; 4. Accounting for direct and indirect greenhouse gas emissions and an emission reductions program; 5. An assessment of energy consumption and an energy efficiency Program; 6. Presence of a biodiversity conservation program in the company's areas of operations; 7. Inclusion of a wildlife rescue section in company's Oil Spill Contingency Plans (OSCPs) and/or Oil Spill Emergency Response Plans (OSERP)

Section 2. Environmental impact
1. Specific gross emissions of air pollutants; 2. Associated Petroleum Gas (APG) utilization rate; 3. Unit volume of polluted water discharged to surface water Bodies; 4. Unit fresh water withdrawal; 5. Ratio of annual waste generation volume to annual waste management volume; 6. Polluted land area ratio at the beginning and end of the reporting year; 7. Disturbed land area ratio at the beginning and end of the reporting year; 8. Share of disturbed and polluted land area to the company's total area of operations; 9. Pipeline leak rate per unit of pipe length; 10. Specific amount of oil, condensate and oil products spilled as result of accidents and leaks; 11. Share of exceptional charges in total charges for adverse environmental impacts; 12. Share of clean fuels (Euro 4-5 high-octane gasoline, Class 3-4-5 diesel, gas motor fuel, and biofuel) in total fuel production; 13. Power generation from renewable energy sources (RES), including that for internal consumption.
Section 3. Disclosure/transparency
1. Non-financial reporting in compliance with GRI requirements; 2. External assurance of non-financial reporting in compliance with GRI requirements; 3. Public access to the Environmental Impact Assessment (EIA) including Internet access, as well as the EIA being available online after public consultations; 4. Public access to OSCPs and OSERP (sections regarding environmental impact mitigation), including guaranteed Internet access; 5. Informing the public of major accidents resulting in significant socio-environmental damage, including those caused by contractor activities; 6. Informing the public on environmental conflicts between the companies and stakeholders in company areas of operations, including conflicts caused by contractor activities; 7. Established procedure for public claim review; 8. Public availability of criteria 1-7 of Section 1 for the reporting period; 9. Public availability of criteria 1-13 of Section 2 for the reporting period.

3.3. Analysis of Rating Results

According to material from (Shvarts et al. 2016, Knizhnikov et al. 2015) authors have made Table 2 which compares of the rating results in 2014 and 2015 years for the russian oil and gas companies, operating in the Arctic.

Table 2. Comparison of the rating results in 2014 and 2015 years for the russian oil and gas companies, operating in in the Arctic [a]

Company	Final 2015 position/ Final 2015 rating point	Point change as compared to 2014 result	Final 2014 position/ Final 2014 rating point
Gazprom[b,c]	1/1.5387	▲ 0.1842	2/1,3545
Sakhalin Energy (Sakhalin-2) [b,c]	2/1.5222	▼-0.0031	1/1,5253
Zarubezhneft[c]	3/1.4222	▲ 0.3666	4/1,0556
LUKOIL[c]	4/1.3737	▲ 0.3794	5/0.9943
Rosneft[b,c]	5/1.3569	▲ 0.2709	3/1,0860
Gazprom Neft[b,c]	6/1.0202	▲ 0.1485	6/0,8717
NOVATEK[c]	7/0.9667	▲ 0.2462	8/0,7205
Bashneft[c]	8/0.9158	▲ 0.1843	7/0,7315
NNK/Alliance[c]	9/0.2828	▲ 0.0976	9/0,2804
RussNeft[c]	10/0.2593	▼-0.0211	10/0,1852
Arcticgas[c]	11/0.1481	-	unrated

[a] Companies were selected based on their volume of production. The lower limit was set at 1.5 million tons of oil equivalent (oil and gas condensate)

[b] Companies operate into Arctic offshore

[c] Companies operate in the AZRF

It is worth mentioning that a number of companies included in table 2 are owners of some of the other companies listed in the same table (e.g., Gazprom – Sakhalin Energy, Gazprom Neft) (Knizhnikov et al. 2015). All the companies in the table are viewed as being independent as each retains its own internal policies as well as being free to formulate corporate social responsibility policy (Shvarts et al. 2015). Some oil and gas companies with large production volumes (e.g., Gazprom, Zarubezhneft) are owned by the government. It's believed these companies deal with weak pressure from civil society. Before oil prices collapsed by 50% in 2014, the oil and gas market was a 'seller's market'. Therefore, improving environmental responsibility standards and transparency in the oil and gas industry through implementing 'soft' responsibility mechanisms is a critical task (Douma 2010, Shvarts et al. 2016) aimed at increase investment attractiveness of russian oil and gas companies and GER's reduction.

The most visible trend of the rating held in 2015 comparing to 2014 is the growth of environmental responsibility and trasparency for most companies (Knizhnikov et al. 2015). The three rating leaders among others companies, operating in the Arctic are Gazprom, Sakhalin Energy and Zarubezhneft. The best positive rating dynamics in 2015 among companies, operating in the Arctic has been demonstrated by LUKOIL, Zarubezhneft, Gasprom and NOVATEK (+1 position year-on-year for each ones). According to (Knizhnikov et al. 2015) these improvements are stipulated, first and foremost, by the companies now disclosing additional information on environmental management and environmental impact. For instance, Zarubezhneft have not only improved information disclosure in their annual non-financial reports, but also made public a large volume of additional information in the Internet sites of their subsidiaries.

In accordance with data from (Knizhnikov et al. 2015) authors created Table 3 that shows final rating points and results of three sections for the oil and gas companies, operating in the Arctic. This table also contains information about different levels of information transparency when it comes to the environmental aspect of their businesses. The most of the companies are maintaining a high level of transparency on environmental issues. This can be explained by the fact that investors pay attention to all company-related information including, naturally, that relating to environmental issues. When a company publishes detailed reports on environmental protection, it decreases the level of public uncertainty related to its business and improves its investment prospects (Shvarts et al. 2016). It is worth noting that the level of transparency amongst some companies is quite low. For example, Russneft and Alliance.

Table 3. Rating 2015 results according to three sections

Company	Section 1. Environmental management		Section 2. Environmental impact		Section 3. Disclosure/ transparency	
	Section 1 position	Rating point	Section 2 position	Rating point	Section 3 position	Rating point
Gazprom[d]	2	1.7778	1	1.7273	4-5	1.1111
Sakhalin Energy (Sakhalin-2)[d]	1	2.0000	5-6	0.9000	1	1.6667
Zarubezhneft[d]	6	1.2222	2	1.6000	2	1.4444
LUKOIL[d]	4	1.5556	3	1.4545	4-5	1.1111
Rosneft[d]	3	1.6667	4	1.1818	3	1.2222
Gazprom Neft[d]	5	1.3333	7	0.7273	6-7	1.0000
NOVATEK[d]	7-8	1.1111	5-6	0.9000	8	0.8889
Bashneft[d]	7-8	1.1111	8	0.6364	6-7	1.0000
NNK/Alliance[e]	10	0.1111	9	0.1818	9	0.5556
RussNeft[e]	9	0.3333	10-11	0.0000	10-11	0.4444
Arcticgas	11	0.0000	10-11	0.0000	10-11	0.4444

[d] High level of information transparency. The company site has a substantially informative chapter specifically dedicated to the company's environmental policy and environmental protection issues. The company publishes sustainable development and/or environmental reports on an annual basis

[e] Low level of information transparency. The company site either completely lacks any chapters dedicated to environmental issues, or the corresponding chapter is not informative enough and contains only general information (at best, the chapter only lists environmental policy principles). The company does not publish sustainable development and/or environmental reports

We can see that leaders in all three sectors are different. The best position of environmental management belong to Sakhalin Energy (Sakhalin-2), Gazprom, Rosneft. In sector 2 "Environmental impact" the main companies, operating in the Arctic, are Gasprom, Zarubezhneft, LUKOIL. Gasprom is one leading state-owned gas producer, which accounts for approximately 70% of total gas production in Russia and approximately 11 large oil extracting companies in the country (Shvarts et al. 2015). It has improved its position from 2014 and now is the leader of

Environmental Impact section, while Sakhalin Energy hasn't undergone any changes. Sector 3 connected with the level of corporate transparency with respect to environmental issues and largely determined by the quality of non-financial reporting. There are three types of such reporting for russian oil and gas companies: environmental reports, social reports, and sustainable development reports. The latter are the most popular form (RSPP 2015; Shvarts et al. 2016). All of the analyzed companies disclose environmental and social parameters of their activity in one way or another. Seven of these 11 companies only publish sustainable development reports (Rosneft, Lukoil, Gazprom Neft, Bashneft, Sakhalin Energy, NOVATEK and Zarubezhneft), and one company publishes reports on sustainable development as well as environmental reports (Gazprom). So the leaders in sector "Disclosure/transparency" are Sakhalin Energy (Sakhalin-2), Zarubezhneft, Rosneft. The best rating dynamics were shown by LUKOIL and Zarubezhneft which also made public a large volume of information in the Internet sites and their subsidiaries (Knizhnikov et al. 2015).

Thus, we can see that the rating of oil and das companies has positive effects in the sphere of environmental improvements and public acknowledgement.

4. Conclusion

Currently with surge in energy demand, the oil and gas industry is being challenged to improve production efficiencies and to maintain a growth trajectory in its crude oil and natural gas resources. The Arctic constitutes an area of new opportunities for oil and gas companies. However, russian oil and gas industry also is faced with barriers to foreign direct investment (FDI) arising from the uncertainty of the existing legal regime of Arctic (geopolitic issues) and the readiness of technologies to cope with geoecological challenges. This underinvestment may lead to oil spills and pipeline leaks that have influence on increasing GER. The introduction of environmental ratings make it possible to prioritize attracting of investments to those russian oil and gas companies, which the most meet the sustainable development criteria.

The rating's results show that there are differences among russian oil and gas companies, operating in the Arctic. It depends on their levels of environmental responsibility and transparency. Large, publicly-traded companies with a strategic focus on gas have received top ratings. The «gas component» of the rating is of particular importance as gas is the "transition" fuel on the path to a fossil-free energy economy (Shvarts et al. 2015, Shvarts 2015). Privately held, smaller oil companies are at the bottom, along with the subsidiaries of major Russian and international corporations (transparency of subsidiary companies tends to be lower than that of parent corporations) (Shvarts et al. 2016). There are «top three» rating companies: Gazprom, Sakhalin Energy (Sakhalin-2), Zarubezhneft.

There are different companies in rating's sectors. The best position of environmental management belong to Sakhalin Energy (Sakhalin-2), Gazprom, Rosneft. In sector "Environmental impact" the main companies are Gasprom, Zarubezhneft, LUKOIL. The leaders in sector "Disclosure/ transparency" are Sakhalin Energy (Sakhalin-2), Zarubezhneft, Rosneft.

Thus, the environmental rating of the russian oil and gas companies, operating in the Arctic, may create a new mechanism for informing foreign investors, which may leads to the rise of FDI in Russia and lowering of the GER in conditions of Arctic's exploitation accordingly.

ACKNOWLEDGMENT

The authors are grateful for comments and information provided by Evgeny Shvarts (Director of Conservation Policy, WWF, Russia) and Alexey Knizhnikov (Head of the Programme for Environment Policy in the Power Generation and Mining Sectors, WWF, Russia).

REFERENCES

Avramenko, A. A., Sanin, S. V. & Vishnyakov, Y. aD. (2006). Problems of rating estimation of investment attractiveness of enterprises based on environmental component. *Journal of Russian entrepreneurship.*, *4* (76), 96-101.

Beurden, P. & Gossling, T. (2008). The worth of values e a literature review on the relation between corporate social and financial performance // *J. Bus. Ethics*, *82*, 407-424.

Bigliani, Roberta. (2013). Reducing Risk in Oil and Gas Operations // White Paper. Sponsored by: EMC. IDC Energy Insights. May 2013.

Diagnostic analysis of the environmental status of the Russian Arctic (Advanced Summary) (2011), Editor-in-Chief: B.A. Morgunov, Moscow, Scientific World, pp. 172.

DJSI, (2015). Review Results. September 2015, RobecoSAM, 32 pp. Available at: http://www.sustainability-indices.com/images/review-presentation-2015.pdf.

Douma, WTh. (2010). The EBRD and Russia: stimulating European principles for the environment. In: Douma, W. Th. & Mucklow, F.M. (Eds.), Environmental Finance and Responsible Business in Russia: Legal and Practical Trends. T.M.C. Asser Press, The Hague, 169-188.

Environmental Management in oil and Gas Exploration and Production (1997). Joint E&P Forum/UNEP Technical Publication. An overview of issues and management approaches. Oxford, United Kingdom. 68 p.

Environmental Standards for Operations of Oil and Gas Companies Acting in Russia, (2005). On its Continental Shelf, and within its Exclusive Economic Zone Developed by Russian Non-governmental Nature Conservation Organizations. Moscow, 8 pp. Avalable at: http:// wwf.ru/resources/publ/book/eng/109.

Global Investment Trade Monitoring. Embargo, Report of United National Conference on Trade and Development, No. 18, 29 January 2015, 17:00 GMT (12:00 New York, 18:00 Geneva).

Improving safety and productivity in oil and gas operations, (2014). // White Paper. Motorola solutions.

Knizhnikov, A. Yu., Pakhalov, A. M., Sipailova, Yu. & Rodionov, K. (2015). Environmental Responsibility Rating Of Oil And Gas Companies In Russia, 22 p.

Official site of "Federal state statistics service" — the flow of foreign investments by kinds of economic activity (2016). Available at: www.gks.ru.

Prakash, A. & Potoski, M. (2012). Voluntary environmental programs: a comparative perspective. *J. Policy Anal. Manag.*, *31*, 123-138.

RSPP, (2015). National Register and Library of Non-financial Reporting (Nacionalniy Registr i Biblioteka korporativnih nefinansovih otchetov). Avalable at: http://www.rspp.ru/simplepage/157.

Sharkey, A. J. & Bromley, P. (2015). Can ratings have indirect effects? Evidence from the organizational response to peers' environmental ratings. *Am. Sociol. Rev.*, *80*, 63-91.

Shvarts, E. A., Bunina, J. & Kniznikov, A. Yu. (2015). Voluntary environmental standards in key russian industries: a comparative analysis // *Int. J. Sus. Dev. Plann.*, Vol. *10*, No. 3, 331–346.

Shvarts, E. A., Pakhalov, A. M. & Kniznikov, A. Yu. (2016). Assessment of environmental responsibility of oil and gas companies in Russia: the rating method //*Journal of Cleaner Production.*, Volume *127*, 20 July 2016, 143–151.

Strizhova, YuS. (2015). Foreign investment in the Russian economy. *Problems of Economics and management.*, *1* (01)., 9-11.

Stuebs, M. & Sun, L. (2010). Corporate governance and environmental performance. *J. Account. Ethics Public Policy*, *11*, 381-395.

The international standard ISO 14001:2015. Environmental management systems – Requirements with guidance for use. The third edition 2015-09-15. Translation By A. Gorbunova, 2015, 46.

Trubitsina, O. P. (2016). Biota monitoring in the impacted zones of oil and gas industry in the arctic region. In: Bashkin, VN (Ed) Biogeochemical Technologies for Managing Environmental Pollution in Polar Ecosystems, NY: Spinger, 87-94.

Trubitsina, O. P. & Bashkin, V. N. (2016). The analysis of geoecological risks and ratings as a factor increasing the investment attractiveness of enterprises. *Issues of Risk Analysis*, vol. *13*. 62-68.

Trubitsina, O. P. & Bashkin, V. N. (2016). The analysis of geoecological risks and ratings as a factor of improving investment attractiveness of enterprises. In: Bashkin, VN (Ed) Biogeochemical Technologies for Managing Environmental Pollution in Polar Ecosystems, NY: Spinger, 141-150.

Trubitsina, O. P. (2014). Investment attractiveness of environmentally oriented enterprises. Actual problems of modern business. The materials of international correspondence scientific-practical conference, Arkhangelsk, 82-85.

Wernstedt, Kris. (2002). Environmental Management in the Russian Federation: A Next Generation Enigma. Resources for the Future, Washington., 37p.

CONCLUSION

Vladimir N. Bashkin*

Institute of Physicochemical and Biological Problems of Soil Sciences of Russian Academy of Sciences, Pushchino, Russian Federation

Institute of Natural Gases and Gas Technologies – Gazprom VNIIGAZ LLC, Razvilka, Russian Federation

Thus, the study of ecological and biogeochemical indicators in gas industry impacted polar ecosystems in the Northern regions, including footprints of climate changes, the simulation of geoecological and geopolitical risk processes and the relevant challenges allows us to simulate the behavior of key elements (nitrogen, carbon) in biogeochemical cycles. This, in turn, can be used in the development and application of biogeochemical technologies, for example, for recultivation of disturbed and polluted areas, calculation of geo-ecological risk indicators and for developing techniques to manage this risk in different geo-ecological situations owing to environmental pollution.

---

* Address correspondence to this author at Institute of physicochemical and biological problems of soil science RAS: Prof., Dr (biol) Vladimir Bashkin, Principal researcher, Pushchino, Moscow region, 142292 Russia. Tel: +7 916 860-2038, Fax: +7 4967 33-0595, e-mail – Bashkin@issp.serpukhov.su

Accordingly, ecological and biogeochemical footprints are shown for possible climate alterations in polar areas of the Yamal peninsula, the Russian Federation. The relative content of soil nitrogen and carbon fractions of soil organic matter can be served as ecological footprints of regional climate changes. The microbial consumption of resistant carbon compounds can be explained by changes in the structure of microbial communities dominated by microorganisms with high temperature optimum, which is capable of utilizing more sustainable carbon substrate. These soil organic matter values may grow during the cyclic periods of regional temperature increase and may decline in the relevant cold periods. The temperature ranking may serve as the biogeochemical footprints of regional climate changes as well. The simulating climate changes in the Yamal peninsula were indicated the modern predominant trends to relevant cooling periods.

Furthermore there were established the regularities of the heterogeneous distribution of reserves of different forms of nitrogen and carbon in soils of tundra ecosystems in space – depending on the meso- and microrelief, and with increasing depth in the soil profile. It was shown that at small scale, the temperature, of course, controls the dynamics of nitrogen in the Arctic ecosystems, which is proved by the higher mean values of the pools of nitrogen and the speed of the cycle in southern tundra compared to the typical. Soil moisture (or the effect of meso-relief) affects the mode of nitrogen in only the soil between bumps along the gradient from temperate-humid to extra-humid conditions. However, in larger scale, the effect of microrelief and the closely related differentiation of vegetation cover the effects of temperature and soil moisture, and acts as the main controlling factor of the nitrogen fund in tundra soils due to the redistribution of litter and the lateral introduction of dissolved organic and mineral substrates in depression zone. The mosaic distribution of nitrogen of the zonal background soils and its rate of spatial mineralization is enhanced by cryoturbation processes that result in party leaching of organo-accumulative horizon with high nitrogen content into the soil depth in conditions of lower temperatures and, accordingly, reduced activity of the microbiota.

This was also shown that the increase of plant biomass to warming in largely due to the response of lichens is the dominant component of the dwarf shrub-lichen tundra. The rate of CO_2 emissions is greatly increased due to warming and leads to a net loss of carbon from the “soil-plant” system at elevated temperatures compared with normal temperatures. This confirms the hypothesis that as a result of climate warming, the net carbon balance of the Arctic ecosystems (at least in the short term) could change from a net sink of carbon to its clean emissions. The responses of individual processes to the warming likely depend on the type of the system under investigation, which requires experimentation with the manipulation of temperature in a variety of plant communities and soil types. However with the change of temperature process on the observed cold is able to restore the status quo in the biogeochemical cycles of carbon and nitrogen including CO_2 emission and sink.

In this volume we paid as well attention to interpreting the ecological interactions in biogeochemical cycles of carbon and nitrogen. The evidence that a stable soil organic matter is preferably decomposed at high temperatures can be extremely important for predicting the impact of climate change on the carbon cycle in the Northern ecosystems. Two points are the most important: a) even the most stable fractions of the soil organic matter can be potentially mineralized by soil microbes in tundra ecosystems; and b) resistant fractions of the soil organic matter can be mineralized in a relatively high degree. Furthermore, the mineralization of resistant carbon compounds can be explained by changes in the structure of microbial communities dominated by microorganisms with high temperature optimum, which is capable of utilizing more sustainable carbon substrate. The results prove that temperature, not the depletion of the pool of readily available carbon compounds lead to a preferred degradation of stable fractions of the soil organic matter. Thus, the inventory of stable carbon in tundra biome may not be permanent. These soil organic matter values may grow during the cyclic periods of regional temperature increase and may decline in the relevant cold periods.

At the base of outdoor chamber method the effect of experimental elevated temperature on tundra ecosystem productivity was studied.

Furthermore, using ^{15}N techniques, it was found that higher vascular plants of tundra ecosystems, in contrast to more southern ecosystems, have developed an adaptive mechanism to the predominant absorption is not nitrate, but ammonia nitrogen. When comparing the quality of nitrogen-containing substrates such as amino group (glycine), nitrate, ammonium and amide (urea), it was found that all plants uptake the minimum quantity of nitrate nitrogen. The speed of assimilation of ammonium nitrogen by vascular plants was higher on average almost 14 times, amide nitrogen 6 times and the nitrogen of the amino group is almost 3 times than nitrate nitrogen. The speed of assimilation of all forms of nitrogen by mosses and lichens were 3-4 times higher than vascular plants. Spore-bearing plants, as well as vascular, intensely absorb more ammonium nitrogen than nitrate. However, in contrast to vascular plants, the rate of assimilation of the nitrogen of the amino group and the amide nitrogen by mosses and lichens is very high.

As regards the using ecological and biogeochemical cycling data for applying in biogeochemical technologies I would like to mention the following. Three new ecologically oriented biogeochemical technologies of reclamation of mechanically disturbed tundra soils of the territories of productive activity of the gas industry in conditions of the Far North are presented. In the first technology the local peat dose depending on a soil full moisture capacity is defined, in the second technology the procedure of receiving from local peat a potassium humate as a stimulator of growth and development of plants is described, and the third technology is devoted to the combined use of local peat and potassium humate at soil reclamation. Accordingly, in this volume the following technologies are presented as the following:

Biogeochemical Technology for Recultivating Disturbed Soils of Island Bely (Kara Sea). Realization of recultivation technology of physically disturbed and coal polluted soils from the Island Bely in the conditions of *in vitro* experiment has shown that there is an possibility of not only rational local peat using but also to proposing remediating means for the *in situ* large-scale and long term soil recultivation to accelerate a restoration of biogeochemical cycles of substances originally at the level of microbe populations, for avoiding full degradation of a soil cover as a basis of existence of native flora and fauna. Such recultivation will be an impulse to the subsequent restoration of various links of biogeochemical cycling at the levels of invertebrate organisms and vegetable biogeocenoses.

Innovative Biogeochemical Technology for Recultivating Disturbed Soils of Taz Peninsula. The data on the biochemical analysis of the researched soils and pot experiment confirm the fact of implementation of innovative biogeochemical technology for recultivating disturbed tundra soils of the Taz Peninsula. In general received experimental data allow to coming to conclusion about further application the technology in the territories where «Gazprom Dobycha Yamburg» LLC carries out exploration works, arrangement and development of new fields, performs gas and gas condensate production and their preparation for transport.

The role of Plant Growth-Promoting Rhizobacteria in the Remediation of Pb-contaminated Soil. A series of pot experiments on soil artificially contaminated with water-soluble lead compounds has investigated the effect of plant growth-promoting *Pseudomonas* rhizobacteria on barley growth, uptake of lead by barley plants and its distribution in soil as a constituent of various compounds. Inoculation with the bacteria significantly reduced the concentration of lead in shoots and roots in the first half of the growing season (up to the heading stage). A decrease in Pb uptake by inoculated plants corresponded to an increased binding of lead in soil in the form of extracted NH_4Oac complexes and organic compounds in the first half of the growing season. Thereby, the mobility of metal was reduced, without changing the reaction of the soil medium, probably due to bacterial production of organic metabolites, such as siderophores. Application of strain *P. fluorescens* 21 eliminated Pb toxicity; herewith, the plant weight (grain weight including) increased up to the level of the uncontaminated uninoculated control. Other possible mechanisms of plant resistance to the toxic effect of lead at the introduction of the bacteria are discussed. The practical application of this approach will allow the recultavation of polluted soil in various climate regimes.

From the geoecological point of view, process of soil reclamation represents the reconstruction of its fertility lost due to various reasons to which it is possible to refer productive activity of the gas industry in the conditions of the Far North. Thus, thc activity of catalase and dehydrogenase enzymes can be considered a tool for controlling soil reclamation process. The relevant results of *in vitro* experiments aimed peat based remediation of the physically disturbed tundra soils of the Taz Peninsula showed that catalase and a dehydrogenase activity analysis is fitting for express assessment of reclamation processes.

In this volume the modern techniques are used for assessing biogeochemical and ecological features of polar ecosystems. As regards NMR ^{13}C measurements for assessing humates from tundra peats, specific task phase of the study was theoretical substantiation and the application of NMR ^{13}C spectroscopy to assess the purity of the preparations of humic substances isolated from peat of tundra ecosystems in the impact area of the Yamburg gas condensate field. To solve this problem 3 local peats were sampled from different peat quarries such as various natural depressions and excavations. The analytical work on the identification of humate structures of isolated compounds was conducted using modern analytical techniques: IR and NMR spectroscopy. The applicability of the method of ^{13}C NMR spectroscopy to assess the preparations of humic substances isolated from peat in the gas industry impacted ecosystems was shown. It is shown that quantitative ^{13}C NMR spectroscopy is one of the most informative methods for physical-chemical studies of the chemical nature of humic acids.

Furthermore, based on the quantitative estimates of biogeochemical cycling in polar ecosystems and the relevant knowledge of their ecological parameters we paid significant attention to assessing risk values and risk management in impacted areas.

As an example, risk management of environmental pollution by methanol as the main inhibitor of hydrate formation in the gas industry in polar region was suggested. The contamination risks to the environment can be shown in the form of emergency emissions or spills in case of production, transportation and use of this substance. Need of control of the environmental pollution by methanol is connected with risk of its toxic action on human and for this purpose using hygienic standard of substance in the form of maximum permissible concentration and maximum permissible level is reasonable. Methods of utilization and cleaning of sewage and soil containing methanol are various (thermal, underground burial, physicochemical, catalytic and microbiological) and which choice for practical application, first of all, will be determined by the most effective environment protection and health of human.

Another example is related to risk management of environmental pollution during oil production and transport. The analysis, systematization and generalization of information concerning a problem of environmental protection against oil contamination are carried out. Examples of the most considerable emergency oil floods to soil and surface waters during its production and transport are presented. The measures for a remediation of the surface waters polluted by oil are described. The special attention is paid to a remediation of soil polluted by oil as a source of its entering in ground waters, by migration on a soil profile. The essence of this remediation consists in soil cleaning by means biocompost for the accelerated microbiological degradation of oil hydrocarbons. Researches showed that entering various doses of biocompost (50 and 100 g/kg) into the soil reduce time of almost full degradation of oil hydrocarbons at concentration in 50 g/kg by 1.8-5.4 times, at concentration of 100 g/kg – 1.7-7.2 times.

It is recognized today, the main risks for the oil and gas industry are risks related to geoenvironment and geopolitics. These risks can be transformed into additional opportunities by a number of ways. For example, this can be done through exploration and development of conventional deposits in a hard-to-reach region such as the Arctic. In this case, expansion of the geographic boundaries of operations opens new opportunities for companies, which have the necessary technologies and resources. However, such development of the Arctic territory involves significant geoenvironmental risks, GERs, and geopolitical risks, GPRs. At the same time, these risks can turn to negative conditions rather than opportunities. This includes the risk of a hybrid war, possible military clashes, etc. Companies should therefore be more flexible in selecting the business strategy and factor in the interrelation and interdependency of the GERs and GPRs. From the global community's standpoint, enhancement of the interaction between geopolitics and geoenvironment appears to be very promising for transformation of GERs and GPRs to additional opportunities. This interaction can have a powerful effect if the concept of sustainable development is emphasized.

In the given volume the sustainable development approach is considered based on ecological and biogeochemical parameters for calculating critical loads of pollutant and their exceedances in polar and sub-polar ecosystems.

The attention is paid to the proposed methodology of assessing polar marine ecosystem risks associated with industrial emissions that not only allows to make quantitative estimation of potential changes in the ecosystem condition but also to calculate probability of their occurrence. It is also providing a detailed characterization of the Barents Sea ecosystems as targets of man-induced impact. Performed assessments of potential risks to marine ecosystems are qualitative in nature, identifying levels of stress, which may cause environmental hazards violations of the trophic structure of the individual sections of the Barents sea. However, in the absence of other available data, they provide the minimum necessary information for making management decisions to reduce potential environmental risks in connection with the development of the fields in shelf zone of the Arctic seas.

Moreover, the integrated dynamic ecosystem models SAFE and ForSAFE-VEG were used to estimate critical loads for nitrogen using different chemical and biological criteria, and attempting to account for possible effects of climate change. The study is based on a model application on a set of 640 sites earlier used for critical loads assessments in Sweden, focusing on the ecological effect of nitrogen pollution. To assess critical loads for nitrogen to ecosystems, two criteria were used for the critical loads estimates: 1: N leaching, and 2: changes in plant community composition, and assessed using the models mentioned. It was shown that the model gives results consistent with earlier results using other methods such as the simple mass balance method. The new model system is operational over the whole of the Swedish territory and may be used for policy development and policy outcome assessments. The critical loads are estimated both under the assumption of no change in climate from today's level and under the assumption that the steady state of climate will be comparable to the predicted levels of 2100 for temperature and precipitation under the IPCC-SRES A2 scenario (Formerly the "worst case" scenario, after 2011, the "medium case" scenario).

As it has been already mentioned the quantification of risk will allow us to manage these risks and environmental ratings as a factor of improving investment attractiveness of Russian oil and gas companies, operating in the Arctic, can be the good example. There were shown the need of using environmental rating for oil and gas industry in Russia. Particular attention is given to the issues of rating results of Environmental Responsibility of Oil and Gas companies in Russia that was first held in 2014 and then in 2015. It was conducted by the cooperative initiative by CREON Group and WWF Russia with participation of National Rating Agency. It was selected from all Russian oil and gas companies only those who operating in the Arctic region and these have been analyzed. The rating's results show that there are "top three" among Russian oil and gas companies, operating in the Arctic. They are Gazprom, Sakhalin Energy (Sakhalin-2), Zarubezhneft. There are also different companies in rating's sectors. For instance, the best positions of environmental management belong to Sakhalin Energy (Sakhalin-2), Gazprom, Rosneft. In whole, the environmental rating of the Russian oil and gas companies, operating in the Arctic, may create a new mechanism for informing foreign investors, which may leads to the rise of FDI in Russia and lowering of the GER (geoenvironmental risk) in conditions of Arctic's exploitation accordingly.

Thus, the studying ecological and biogeochemical cycling in impacted polar ecosystems enhanced the developing biogeochemical technologies for recaltivation of polluted and disturbed areas as well as the quantitative methods for relevant risk assessing and managing.

About the Editor

Vladimir N. Bashkin graduated from biology-soil faculty of M. V. Lomonosov Moscow state University in 1971. In 1975 he defended a PhD at MSU, and in 1987 - his doctoral thesis. In 1997 he was awarded the academic title "Professor" in the specialty "biogeochemistry". Since 1971 he works at the Pushchino Scientific Center of Biological research of the Russian Academy of Sciences. For several years he taught in different universities of the world, in particular, at Cornell University, USA, Seoul National University, Korea, Royal University of Technology Bangkok, Thailand, M. V. Lomonosov Moscow state University, Russia. Currently, he is principal researcher of the Institute of physicochemical and biological problems of soil science of RAS, also part-time chief scientific officer of LLC "Gazprom VNIIGAZ". Scientific interests of Professor V. N. Bashkin include biogeochemistry, systems analysis, geo-ecological risk of transboundary transfer of polluting substances, the remediation of contaminated territories. Great attention is paid to the development of innovative biogeochemical technologies. V. N. Baskin published more than 40 books in English, Chinese and Russian domestic and foreign publishing houses, he is the author and coauthor of more than 300 scientific articles published in leading scientific journals, received 16 patents. He has supervised 5 doctoral and 20 PhD theses in different countries of the world. Prof. V. N. Bashkin is member of the editorial Board of 5 international journals. For several years he was the Deputy

Chairman of the Working Group on Effects (scientific Committee) of the UN Long-Range Transboundary Air Pollution Convention.

INDEX

#

A

B

C

D

E

F

G

H

I

L

M

N

O

P

R

S

T